Organic-Inorganic Hybrid Materials

Organic-Inorganic Hybrid Materials

Editors

Jesús-María García-Martínez
Emilia P. Collar

MDPI • Basel • Beijing • Wuhan • Barcelona • Belgrade • Manchester • Tokyo • Cluj • Tianjin

Editors
Jesús-María García-Martínez
Polymer Engineering Group (GIP)
Polymer Science and Technology Institute (ICTP)
Spanish Council for Scientific Research (CSIC)
Spain

Emilia P. Collar
Polymer Engineering Group (GIP)
Polymer Science and Technology Institute (ICTP)
Spanish Council for Scientific Research (CSIC)
Spain

Editorial Office
MDPI
St. Alban-Anlage 66
4052 Basel, Switzerland

This is a reprint of articles from the Special Issue published online in the open access journal *Polymers* (ISSN 2073-4360) (available at: http://www.mdpi.com).

For citation purposes, cite each article independently as indicated on the article page online and as indicated below:

LastName, A.A.; LastName, B.B.; LastName, C.C. Article Title. *Journal Name* **Year**, *Volume Number*, Page Range.

ISBN 978-3-0365-1301-0 (Hbk)
ISBN 978-3-0365-1302-7 (PDF)

© 2021 by the authors. Articles in this book are Open Access and distributed under the Creative Commons Attribution (CC BY) license, which allows users to download, copy and build upon published articles, as long as the author and publisher are properly credited, which ensures maximum dissemination and a wider impact of our publications.

The book as a whole is distributed by MDPI under the terms and conditions of the Creative Commons license CC BY-NC-ND.

Contents

About the Editors vii

Jesús-María García-Martínez and Emilia P. Collar
Organic–Inorganic Hybrid Materials
Reprinted from: *Polymers* **2021**, *13*, 86, doi:10.3390/polym13010086 1

Coro Echeverría, Miguel Rubio and Daniel López
Thermo-Reversible Hybrid Gels Formed from the Combination of Isotactic Polystyrene and [Fe(II) (4-Octadecyl-1,2,4-Triazole)$_3$(ClO$_4$)$_2$]$_n$ Metallo-Organic Polymer: Thermal and Viscoelastic Properties
Reprinted from: *Polymers* **2019**, *11*, 957, doi:10.3390/polym11060957 5

Wei Peng, Ying Qian, Tong Zhou, Shenglin Yang, Junhong Jin and Guang Li
Influence of Incorporated Polydimethylsiloxane on Properties of PA66 Fiber and Its Fabric Performance
Reprinted from: *Polymers* **2019**, *11*, 1735, doi:10.3390/polym11111735 21

Xinxin Sheng, Sihao Li, Yanfeng Zhao, Dongsheng Zhai, Li Zhang and Xiang Lu
Synergistic Effects of Two-Dimensional MXene and Ammonium Polyphosphate on Enhancing the Fire Safety of Polyvinyl Alcohol Composite Aerogels
Reprinted from: *Polymers* **2019**, *11*, 1964, doi:10.3390/polym11121964 31

Rosa Barranco-García, José M. Gómez-Elvira, Jorge A. Ressia, Lidia Quinzani, Enrique M. Vallés, Ernesto Pérez and María L. Cerrada
Variation of Ultimate Properties in Extruded iPP-Mesoporous Silica Nanocomposites by Effect of iPP Confinement within the Mesostructures
Reprinted from: *Polymers* **2020**, *12*, 70, doi:10.3390/polym12010070 49

Rui P. C. L. Sousa, Bárbara Ferreira, Miguel Azenha, Susana P. G. Costa, Carlos J. R. Silva and Rita B. Figueira
PDMS Based Hybrid Sol-Gel Materials for Sensing Applications in Alkaline Environments: Synthesis and Characterization
Reprinted from: *Polymers* **2020**, *12*, 371, doi:10.3390/polym12020371 71

Adrián Leonés, Alicia Mujica-Garcia, Marina Patricia Arrieta, Valentina Salaris, Daniel Lopez, José Maria Kenny and Laura Peponi
Organic and Inorganic PCL-Based Electrospun Fibers
Reprinted from: *Polymers* **2020**, *12*, 1325, doi:10.3390/polym12061325 89

Silvia Vita, Rico Ricotti, Andrea Dodero, Silvia Vicini, Per Borchardt, Emiliano Pinori and Maila Castellano
Rheological, Mechanical and Morphological Characterization of Fillers in the Nautical Field: The Role of Dispersing Agents on Composite Materials
Reprinted from: *Polymers* **2020**, *12*, 1339, doi:10.3390/polym12061339 105

Nantikan Phuhiangpa, Worachai Ponloa, Saree Phongphanphanee and Wirasak Smitthipong
Performance of Nano- and Microcalcium Carbonate in Uncrosslinked Natural Rubber Composites: New Results of Structure–Properties Relationship
Reprinted from: *Polymers* **2020**, *12*, 2002, doi:10.3390/polym12092002 117

Bàrbara Micó-Vicent, Marina Ramos, Francesca Luzi, Franco Dominici, Valentín Viqueira, Luigi Torre, Alfonso Jiménez, Debora Puglia and María Carmen Garrigós
Effect of Chlorophyll Hybrid Nanopigments from Broccoli Waste on Thermomechanical and Colour Behaviour of Polyester-Based Bionanocomposites
Reprinted from: *Polymers* **2020**, *12*, 2508, doi:10.3390/polym12112508 . . . . **133**

Jesús-María García-Martínez and Emilia P. Collar
On the Combined Effect of Both the Reinforcement and a Waste Based Interfacial Modifier on the Matrix Glass Transition in iPP/a-PP-*p*PBMA/Mica Composites
Reprinted from: *Polymers* **2020**, *12*, 2606, doi:10.3390/polym12112606 . . . . **153**

Yuxia Liang, Xiaonan Huang, Lanzhong Yao, Ru Xia, Ming Cao, Qianqian Ge, Weibin Zhou, Jiasheng Qian, Jibin Miao and Bin Wu
Regulation of Polyvinyl Alcohol/Sulfonated Nano-TiO_2 Hybrid Membranes Interface Promotes Diffusion Dialysis
Reprinted from: *Polymers* **2021**, *13*, 14, doi:10.3390/polym13010014 . . . . **167**

About the Editors

Jesús-María García-Martínez holds a Ph.D. in Chemistry (Chemical Engineering) by the Universidad Complutense de Madrid (1995). Additionally, he holds two M.Sc. level degrees and has completed more than 70 highly specialized courses in industrial chemistry; polymer science and technology; quantitative research; environmental science; materials science and characterization; R+D+i management, modelling of processes; quality systems in R+D+i, educational sciences, philosophy of science, and so on. He is a Tenured Scientist at the Institute of Polymer Science and Technology (ICTP) of the Spanish National Research Council (CSIC). Since 1992, within the Polymer Engineering Group (GIP), he has co-authored more than 175 scientific and/or technical works in the topics related to polymer engineering, chemical modification of polymers, heterogeneous materials based on polymers, interphases and interfaces, polymer composites blends and alloys, organic-inorganic hybrids materials, polymer recycling, quality, and so on. Dr. García-Martínez has participated in more than 31 research projects financed with either public (national and international programs) or industrial funds, and is co-author of one currently active industrial patent on polymer recycling. As an additional task to his scientific research, in the period 2000–2005, he assumed the position of Quality Director for the ICTP (CSIC) ISO 17025 Accreditation Project of the ICTP Laboratories (ACiTP) under the auspices of the ICTP (CSIC) Directorate. Additionally, Dr. García-Martínez plays a very active role in reviewing tasks for WOS and SCOPUS indexed journals, with more than 250 reports in the last years. In fact, he has been awarded twice with the Publons Reviewer Awards (2018, and 2019), and once with the POLYMERS Outstanding Reviewer Award (2019) for these activities. Since 2016, he is the Head of the Department of Chemistry and Properties of Polymer Materials within the ICTP (CSIC).

Emilia P. Collar, Ph.D. in Industrial Chemistry (U. Complutense, 1986), joined in 1990 at the permanent staff at the Consejo Superior de Investigaciones Científicas (CSIC), serving as a Tenured Scientist, after two-years (1986–88) as CSIC's postdoc fellow, and one more (1989) as Chemical Engineering Assistant Teacher at the Universidad Complutense de Madrid. At the Polymer Science and Technology Institute (Instituto de Ciencia y Tecnología de Polímeros), ICTP/CSIC, she works at the Polymer Engineering Group (GIP), founded in 1982 by Prof. O. Laguna, being the GIP's Head from 1999. Between 1990 and 2005, she supervised six doctoral theses and five post-graduate ones, developed under 14 research projects (public funds supported), as well as 18 research contracts (private funds supported), issuing 53 technical reports for different companies. From 2001 to 2005, she performed the positions of Technical Director of the Physical and Mechanical Properties Laboratory, LFM, and Deputy-Technical Director of the Thermal Properties Laboratory, LPT, under the successful ISO 17025 Accreditation Project for the CSIC/ICTP's Laboratories, ACiTP, commanded by the ICTP's Head. Author and co-author of more than twenty chapters on books, two currently active industrial patents on polymer recycling, and more than one hundred and fifty research papers mostly on SCI Journals, her research work lines deal with polymers and environment under the general frame of heterogeneous materials based on polymers (*Polymers and Environment; Polymer Recycling; Polymer Blends; Composites; Interface Optimization; Interfacial Agents; Chemical Modification of Polymers; Blend Technologies; Morphology and Modeling*). From 2006 to date, she has participated in three Spanish public funded research projects and one EU public funded project under its 7th Framework Program.

Editorial
Organic–Inorganic Hybrid Materials

Jesús-María García-Martínez * and Emilia P. Collar *

Polymer Engineering Group (GIP), Polymer Science and Technology Institute (ICTP),
Spanish Council for Scientific Research (CSIC), C/Juan de la Cierva, 3, 28006 Madrid, Spain
* Correspondence: jesus.maria@ictp.csic.es (J.-M.G.-M.); ecollar@ictp.csic.es (E.P.C.)

Citation: García-Martínez, J.-M.;
Collar, E.P. Organic–Inorganic Hybrid
Materials. *Polymers* **2021**, *13*, 86.
https://doi.org/10.3390/
polym13010086

Received: 24 December 2020
Accepted: 25 December 2020
Published: 28 December 2020

Publisher's Note: MDPI stays neutral with regard to jurisdictional claims in published maps and institutional affiliations.

Copyright: © 2020 by the authors. Licensee MDPI, Basel, Switzerland. This article is an open access article distributed under the terms and conditions of the Creative Commons Attribution (CC BY) license (https://creativecommons.org/licenses/by/4.0/).

According to the IUPAC (International Union of Pure and Applied Chemistry), a hybrid material is that composed of an intimate mixture of inorganic components, organic components, or both types of components which usually interpenetrate on scales of less than 1 µm [1]. This definition matches the subject of all the articles published in this special issue. Thus, the concept of organic–inorganic hybrid materials can be applied to a broad number of approaches. In the same way, the so-called organic–inorganic materials can be revealed as multi-component compounds having at least one of their organic (the polymer) or inorganic component in the sub-micrometric and, more usually, in the nano-metric size domain [2]. Further, the organic–inorganic hybrid systems can be classified into two classes named class I and Class II, depending on if the interaction of the phases is weak or strong, respectively [2]. The latter depends on the type of interactions: Van der Waals, hydrogen, electrostatic, and other interactions for Class I, and due to real chemical bonds between phases (Class II), the coexistence of both types of interactions (I and II) in the same system is possible [2]. In such a scenario, this issue includes a wide range of topics covering these definitions. Consequently, the organic–inorganic materials are revealed as multi-component compounds having at least one of their organic (the polymer) or inorganic component in the nano-metric size domain, and usually offers much better performance than their non-hybrid counterparts [2–6].

Under the above-mentioned premises, this special issue compiles 11 articles written by experts in the field and are devoted to the very different approaches coexisting in the field.

In this context, the first article by Echeverría, Rubio, and López [7] studies of thermo-reversible organic–inorganic hybrid gels by means of thermal and viscolestic properties. For such purposes, they used synthetized hybrid thermo-reversible gels by combining a metallo-organic polymer and isotactic polystyrene with the purpose of solving the poor mechanical properties of these types of compounds. They were able to determine three transitions upon heating (monotectic transition of the iPS gel, melting of the iPS gel, and melting of the metal-organic polymer gel) suggesting that the two polymers are formed independently in the hybrid gel, concluding that these hybrid gels are interpenetrated organic/metallo-organic networks.

Li et al. [8] investigated the influence of incorporating polydimethylsiloxane on the properties of PA66 fiber fabrics, obtained by melt blend spinning, by producing fibers with excellent mechanical properties and reduced hot shrinkage, without remarkable effect on crystallization and melting behavior of PA66. Thus, the effect of the additive on the surface tension suggests the antifouling, waterproof, and stain-resistant ability of the fiber and its fabric. These results are compared with silicon-coated fibers, resulting in the proposition by the authors of the compounds having many properties besides being ecofriendly.

Related to the hot topic of fire resistance materials, Sheng, Lu, and co-workers report a study about it [9]. So, this article is devoted to fire and smoke suppressions of polyvinyl alcohol (PVA) aerogels in order to avoid the fire hazard that they present. For this purpose, the authors used a 2D transition-metal carbide for enhancing the flame-retardant performance of PVA/ammonium polyphosphate-based composites. The results demonstrated that the presence of the carbide in low amounts can boost the flame retardance of the composites even as to pass as V-0 rated. Furthermore, the authors demonstrate that the presence of the carbide greatly reduces the release of heat and smoke, proving the synergistic effect of the ammonium phosphate and the carbide used for the objective of the study.

An investigation about the changes observed on the ultimate properties of isopropylene–mesoporous silica nanocomposites obtained by extrusion caused by the effect of the polymers confinement within the mesostructures is the subject of the article by Cerrada and working group [10]. The study concludes the key role of pore size in the confinement of iPP chains within the nanometric mesostructures of the silicas used, affecting the thermal stability and the ultimate properties of the materials. The main conclusions of the work were the related effect of the silica on the polypropylene observed through the degradation and the rheological behavior of the composites, which is explained by variations in the inclusion of the polymer chains within the mesostructures of the silicas used. The latter was demonstrated by synchrotron SAXS measurements.

Since the carbonation of concrete is a prime problem in civil engineering, the paper by Sousa, Ferreira, and co-workers [11] is related to the reduction of this concrete degradation by the effect of organic–inorganic hybrid films for the potential functionalization of optical fiber sensors to be applied in concrete structures. Virgin films and those doped with phenolphthalein were characterized by a series of techniques and the results obtained were highly promising regarding the properties under high alkaline scenarios causing concrete carbonation.

By considering the increasing importance of additive processing, the work by Peponi et al. [12] is focused on the fabrication of organic and inorganic poly(e-caprolactone)-based electrospun fibers in order to obtain different nanocomposite mats based on poly(e-caprolactone), PCL, reinforced with a series of both organic and inorganic nanoparticles (just 1%). As organic particles, the authors used cellulose nanocrystals (produced by them) and commercial chitosan and graphene, while the inorganic ones were silver and hydroxyapatite particles. The authors found no appreciable differences in terms of thermal parameters. The authors conclude that while the flexibility of the mats increases for whatever type of particle incorporated, the stiffness of the system increases only in the case of silver or graphene particle is present in the mat.

The effect of the dispersing agents with an ionic or steric action on the interactions between hollow glass microspheres and an epoxy-polyamide resin on composite materials for nautical applications are studied by Dodero, Vicini, and colleagues [13]. The composites are evaluated by means of the rheology of the un-crosslinked material as well as mechanical and morphological properties of the crosslinked composites. The authors conclude that steric dispersing agents offer much better compatibility than ionic ones based on their hindrance capability, and so a better-performing composite with a less-marked Payne effect is obtained. The authors claim to be the pioneers in studying the role of dispersing agents for this type of material in comparison to the merely empirical approaches followed for the preparation of fillers for this type of composite.

Smitthipong and coworkers [14] studied the performance of nano- and micro-calcium carbonate fillers in the final properties of un-crosslinked natural rubber compounds, by paying attention the high cost reduction for industry and the half reinforcing character of this type of fillers promoting better mechanical properties. The authors found a similar tendency of results with the independence of the nano or micro character of the nano-filler used; they also show the importance and influence of the specific surface area. However, the authors conclude the effect of the specific surface area on the final properties, as well as the theoretical models which agree with the results. Additionally, they showed (through time–temperature superposition master curves) where the higher filled composites exhibited lower free volume. Finally, they conclude about the real possibility of using nano-calcium carbonate material for the production of tailor rubber-based products.

Since bio-based nanocomposites have emerged as a hot topic, Garrigos, Puglia, and colleagues [15] have studied the effect of chlorophyll hybrid nanopigments obtained by combining chlorophyll dye (obtained from agro-food wastes), calcined hydrotalcite, and montmorillonite nanoclay, on the final color and thermomechanical properties, by using statistical design of experiments, DOEs. In this way, the authors were capable of identifying the more important parameters influencing the mechanical, thermal, structural, morphological, and color properties of the bionanocomposites studied. Hence, the potential use of green natural dyes from broccoli wastes, adsorbed into nanoclays, have been proven to be very interesting in the development of naturally colored bionanocomposites.

Attending to the fact that the interphase between the organic and the inorganic components of the hybrid material is what governs their ultimate behavior, García-Martínez and Collar [16] investigated the combined effect of the reinforcement and the interfacial agent in a polypropylene-based composite. In the article, the authors studied mica platelets and a polymer waste interfacial modifier consisting of a grafted atactic polypropylene with attached p-phenylen-bis-maleamic acid polar groups, by paying attention to the changes of the glass transition temperature. By using Box–Wilson experimental designs, the article focuses on the prediction of how this parameter changes as a function of both the interfacial agent and the nano-reinforcement by taking into consideration the complex character of the iPP/aPP-*p*PBMA/mica system, wherein the interaction between the components will define the final behavior, and that the glass transition is a design threshold for the ultimate properties of parts based in this type of organic–inorganic hybrid materials. Hence, the final purpose of the work is the prediction and interpretation of the effect of both variables on this key parameter, being identified the same critical points as by other properties such as flexural, tensile, impact, and thermal properties, as determined previously by the authors [16].

Finally, by focusing the attention on the fact that is the adjustment of the organic–inorganic interfacial chemical environment which plays a key role for the separation performance of composite materials, Miau, Wu et al. [17] focused their attention on the regulation about how a polyvinyl alcohol/sulfonated nano-TiO_2 hybrid membranes interface promotes diffusion dialysis. For such purposes, the authors investigated the behavior of a series of hybrid membranes fabricated by combining polyvinyl alcohol (PVA) and sulfonated nano-TiO_2 (SNT). In fact, they regulate the effects of the interfacial chemical surroundings on ion transfer as a function of the SNT content to affect the dialysis coefficients of both the hydroxyl ions and the separation factors in order to be of an adequate interval as to make the sulfonic groups in the interfacial regions and the hydroxyl groups present in the PVA main chain. This can have important roles during the transport of sodium and hydroxyl ions. Moreover, the authors identified a membrane with optimal performance in the case of 3% of SNT.

Heterogeneous materials based on organic polymers (as a whole or as a part) can play an essential roles in the fight against COVID-19. In fact, they can be key materials for fabricating preventive and isolation clothing for the medical teams in close contact with ill patients, for life support equipment, containment, and transport thereof. Additionally, organic–inorganic hybrid materials remain the realm for other diverse applications such as controlled dosing of drugs, membrane systems, multi-component packaging, and a myriad of accessories and sanitary items in general. The challenge of continually attending to new demands remains. The latter constitutes the prime motivation and concern for hybrid materials research. This extensively applies to the works compiled in this special issue.

We thank all the authors for their contributions and encourage them to pursue with their research.

Funding: This research received no external funding.

Conflicts of Interest: The authors declare no conflict of interest.

References

1. Alemán, J.; Chadwick, A.V.; He, J.; Hess, M.; Horie, K.; Jones, R.G.; Kratochvíl, P.; Meisel, I.; Mita, I.; Moad, G.; et al. Definitions of terms relating to the structure and processing of sols, gels, networks, and inorganic–organic hybrid materials (IUPAC recommendations 2007). *Pure Appl. Chem.* **2007**, *79*, 1801–1829. [CrossRef]
2. Faustini, M.; Nicole, L.; Ruiz-Hitzky, E.; Sanchez, C. History of Organic-Inorganic Hybrid Materials: Prehistory, Art, Science, and Advanced Applications. *Adv. Funct. Mater.* **2018**, *28*, 1704158. [CrossRef]
3. Pielichowski, K.; Majka, T.M. *Polymer Composites with Functionalized Nanoparticles: Synthesis, Properties, and Applications*; Elsevier Inc.: Amsterdam, The Netherlands, 2019; pp. 1–504.
4. Pogrebnjak, A.D.; Beresnev, V.M. *Nanocoatings Nanosystems Nanotechnologies*; Bentham Books: Sharjah, UAE, 2012. [CrossRef]
5. Lazzara, G.; Cavallaro, G.; Panchal, A.; Fakhrullin, R.; Stavitskaya, A.; Vinokurov, V.; Lvov, Y. An assembly of organic-inorganic composites using halloysite clay nanotubes. *Curr. Opin. Colloid Interface Sci.* **2018**, *35*, 42–50. [CrossRef]
6. Collar, E.P.; Areso, S.; Taranco, J.; García-Martínez, J.M. Heterogeneous Materials based on Polypropylene. In *Polyolefin Blends*, 1st ed.; Nwabunma, D., Kyu, T., Eds.; Wiley-Interscience: Hoboken, NJ, USA, 2008; pp. 379–410.
7. Echeverría, C.; Rubio, M.; López, D. Thermo-Reversible Hybrid Gels Formed from the Combination of Isotactic Polystyrene and [Fe(II) (4-Octadecyl-1,2,4-Triazole)$_3$(ClO$_4$)$_2$]$_n$ Metallo-Organic Polymer: Thermal and Viscoelastic Properties. *Polymers* **2019**, *11*, 957. [CrossRef] [PubMed]
8. Peng, W.; Qian, Y.; Zhou, T.; Yang, S.; Jin, J.; Li, G. Influence of Incorporated Polydimethylsiloxane on Properties of PA66 Fiber and Its Fabric Performance. *Polymers* **2019**, *11*, 1735. [CrossRef] [PubMed]
9. Sheng, X.; Li, S.; Zhao, Y.; Zhai, D.; Zhang, L.; Lu, X. Synergistic Effects of Two-Dimensional MXene and Ammonium Polyphosphate on Enhancing the Fire Safety of Polyvinyl Alcohol Composite Aerogels. *Polymers* **2019**, *11*, 1964. [CrossRef] [PubMed]
10. Barranco-García, R.; Gómez-Elvira, J.M.; Ressia, J.A.; Quinzani, L.; Vallés, E.M.; Pérez, E.; Cerrada, M.L. Variation of Ultimate Properties in Extruded iPP-Mesoporous Silica Nanocomposites by Effect of iPP Confinement within the Mesostructures. *Polymers* **2020**, *12*, 70. [CrossRef] [PubMed]
11. Sousa, R.P.C.L.; Ferreira, B.; Azenha, M.; Costa, S.P.G.; Silva, C.J.R.; Figueira, R.B. PDMS Based Hybrid Sol-Gel Materials for Sensing Applications in Alkaline Environments: Synthesis and Characterization. *Polymers* **2020**, *12*, 371. [CrossRef] [PubMed]
12. Leonés, A.; Mujica-Garcia, A.; Arrieta, M.P.; Salaris, V.; Lopez, D.; Kenny, J.M.; Peponi, L. Organic and Inorganic PCL-Based Electrospun Fibers. *Polymers* **2020**, *12*, 1325. [CrossRef] [PubMed]
13. Vita, S.; Ricotti, R.; Dodero, A.; Vicini, S.; Borchardt, P.; Pinori, E.; Castellano, M. Rheological, Mechanical and Morphological Characterization of Fillers in the Nautical Field: The Role of Dispersing Agents on Composite Materials. *Polymers* **2020**, *12*, 1339. [CrossRef] [PubMed]
14. Phuhiangpa, N.; Ponloa, W.; Phongphanphanee, S.; Smitthipong, W. Performance of Nano- and Microcalcium Carbonate in Uncrosslinked Natural Rubber Composites: New Results of Structure–Properties Relationship. *Polymers* **2020**, *12*, 2002. [CrossRef] [PubMed]
15. Micó-Vicent, B.; Ramos, M.; Luzi, F.; Dominici, F.; Viqueira, V.; Torre, L.; Jiménez, A.; Puglia, D.; Garrigós, M.C. Effect of Chlorophyll Hybrid Nanopigments from Broccoli Waste on Thermomechanical and Colour Behaviour of Polyester-Based Bionanocomposites. *Polymers* **2020**, *12*, 2508. [CrossRef] [PubMed]
16. García-Martínez, J.M.; P Collar, E. On the Combined Effect of Both the Reinforcement and a Waste Based Interfacial Modifier on the Matrix Glass Transition in iPP/a-PP-pPBMA/Mica Composites. *Polymers* **2020**, *12*, 2606. [CrossRef] [PubMed]
17. Liang, Y.; Huang, X.; Yao, L.; Xia, R.; Cao, M.; Ge, Q.; Zhou, W.; Qian, J.; Miao, J.; Wu, B. Regulation of Polyvinyl Alcohol/Sulfonated Nano-TiO$_2$ Hybrid Membranes Interface Promotes Diffusion Dialysis. *Polymers* **2021**, *13*, 14. [CrossRef]

Article

Thermo-Reversible Hybrid Gels Formed from the Combination of Isotactic Polystyrene and [Fe(II) (4-Octadecyl-1,2,4-Triazole)$_3$(ClO$_4$)$_2$]$_n$ Metallo-Organic Polymer: Thermal and Viscoelastic Properties

Coro Echeverría *, Miguel Rubio and Daniel López *

Instituto de Ciencia y Tecnología de Polímeros (ICTP-CSIC), C/Juan de la Cierva 3, 28006 Madrid, Spain; sambaya.music@gmail.com
* Correspondence: cecheverria@ictp.csic.es (C.E.); daniel.l.g@csic.es (D.L.)

Received: 29 April 2019; Accepted: 29 May 2019; Published: 1 June 2019

Abstract: Nano-sized one-dimensional metallo-organic polymers, characterized by the phenomenon of spin transition, are excellent candidates for advanced technological applications such as optical sensors, storage, and information processing devices. However, the main drawback of this type of polymers is their fragile mechanical properties, which hinders its processing and handling, and makes their practical use unfeasible. To overcome this problem, in this work, hybrid thermo-reversible gels are synthesized by combination of a metallo-organic polymer and isotactic polystyrene (iPS) in cis-decaline. A detailed investigation of the thermal and viscoelastic properties of the hybrid gels, in terms of iPS and metallo-organic polymer concentration is performed by means of differential scanning calorimetry and oscillatory rheology, respectively. From the analysis of the thermal properties, three transitions have been determined upon heating: Monotectic transition of the iPS gel, melting of the iPS gel, and melting of the metal-organic polymer gel, which suggest that the gels of the two polymers are formed independently in the hybrid gel, as long as the two polymers are in concentrations above the corresponding critical gelation concentrations. Results regarding viscoelastic properties and morphology confirmed that hybrid gels consisted of an interpenetrated network of polymer gels, formed by iPS and metallo-organic poymer gels growing independently.

Keywords: metallo-organic polymer; isotactic polystyrene; hybrid gel; thermo-reversible gel

1. Introduction

Nano-sized one-dimensional metallo-organic polymers have demonstrated to be excellent candidates for advanced technological applications such as optical sensors, storage, and information processing devices, etc. [1–6]. In particular, polymers of the transition metal ions family with d4–d7 electronic configuration and octahedral symmetry constitute a group of new functional materials with very interesting optical, magnetic, and electronic properties [1,2]. These polymers are characterized by the phenomenon of spin transition, as the polymer can present two electronic states (corresponding to two magnetic states) switchable by the effect of temperature, pressure, light, etc. [6,7]. The change from one state to another is usually accompanied by changes in the physical properties of the material, such as changes in color, dielectric constant, etc., which confer additional applications to these materials [8].

However, these polymers present serious problems such as fragility from the mechanical point of view, instability to oxidation, the difficulty of processing and handling, which make their practical use unfeasible. Therefore, the main challenge of these systems for the development of their potential applications is how to transfer the inherent properties of the polymer in solid state to systems suitable for technological applications. For certain applications, a specific spatial arrangement is crucial. Research performed over the last decades proposed the preparation of SCO materials as thin-films [9,10], liquid crystals [11], and as supramolecular gels [12,13]. In order to overcome this problem, in our previous study, electrospun fibers of blends of the [Fe(II) (4-octadecyl-1,2,4-triazole)$_3$(ClO$_4$)$_2$]$_n$ metallo-organic polymer and atactic polystyrene were developed. [14] This approach allowed to obtain fibers showing the mechanical properties of the organic polymer (atatic polystyrene) but retaining the magnetic properties of the pure metallo-organic polymer [14]. Prior to that work, we reported the development of [Fe(II) (4-octadecyl-1,2,4-triazole)$_3$(ClO$_4$)$_2$]$_n$ metallo-organic polymer gel and the effect of the solvent on the gelation properties [15]. We demonstrated, by means of Small Angle Neutron Scattering (SANS) and rheology, that the responsibility for the gelation is with the side-by-side aggregation into fibers of the individual metallo-organic polymer chains [13]. The formation of hybrid thermo-reversible gels from organic and metallo-organic polymers is also postulated as an interesting technological alternative. The obtained hybrid materials would fulfil the mechanical requirements due to the presence of the organic polymer, and open a new field of applications with the additional functionalities of the metallo-organic polymer [16]. For the fabrication of this kind of materials, it is necessary to find an appropriate solvent in which the polymers can form gels, or at least one of the polymers form a gel and induce the gelation of the second polymer. For instance, Dasgupta et al. [17] developed a study in which hybrid thermo-reversible gels are formed from organic polymers (atactic, isotactic, and syndiotactic polystyrene) and organogels (prepared from trans-oligo(p-phenylenvynilene) molecules) in organic non-polar solvents: cis-decalin, trans-decalin, and benzene. The obtained hybrid gel is composed of two fiber-like networks, which correspond to each polymer growing independently.

Therefore, in this work, hybrid thermo-reversible gels are synthesized by combination of a metallo-organic polymer and isotactic polystyrene in cis-decaline. In this case, cis-decaline is the solvent in which both polymers form gels. In addition, the viscoelastic properties and the morphology of thermo-reversible hybrid gels are also studied in order to determine the compatibility of the two gels as well as the properties of the resulting system.

2. Materials and Methods

Materials. For the synthesis of 4-octadecyl-1,2,4-triazol, monoformylhydrazine (Aldrich, St. Quentin Fallavier, France), triethylorthoformate ((Fluka, Seelze, Germany), octadecylamine (Fluka), and 1-pentanol (Aldrich) were used. Monoformylhydrazine and 1-pentanol were dried before use. For the synthesis of the iron complex Iron (II), Perchlorate (Aldrich) and ethanol (Merck, Kenilworth, NJ, USA) were used without previous purification. Isotactic polystyrene (90% isotactic) (iPS) PS2 (Rapra Technology, Akron, OH, USA), an average molecular weight of 136,000 g/mol and a polydispersity index of 3.8, was used. Cis-decalina (Aldrich) was used as solvent for the gel preparation.

The metallo-organic polymer [Fe(II) (4-octadecyl-1,2,4-triazole)$_3$(ClO$_4$)$_2$]$_n$ was obtained by a method reported previously [15]. Gels of metallo-organic polymer were obtained by dissolving the polymer in cis-decaline at 100 °C. Then, the obtained solution was cooled down to 0 °C in an iced bath until the gel is formed. In the case of iPS pure gels, the preparation was similar except for the temperature applied. An appropriate iPS content was dissolved in cis-decaline at 180 °C. The homogeneous solutions were then cooled down to 0 °C in an ice bath. In both cases, the cooling process needs to be fast enough to prevent polymer crystallization. Hybrid gels were obtained from the preparation of independent homogeneous solution of both polymers in cis-decaline at 100 °C for the metallo-organic polymer and 180 °C for the solution of iPS. Then, both solutions were mixed and stirred at 100 °C to form a homogeneous mixture that was further cooled down to 0 °C to obtain the gel.

For this study, pure iPS gels with concentrations of polymer in the range of 2% to 9% (m/v) were prepared in cis-decaline. For the hybrid gels, two different sets of samples were prepared in cis-decaline: (i) Hybrid gels with a fixed concentration of iPS (4% (w/v)) and metallo-organic polymer concentration in the range of 0.1% to 6% (m/v); (ii) hybrid gels with a fixed concentration of metallo-organic polymer (1.4% (w/v)) and iPS concentrations in the range of 1% to 7% (m/v).

Methods. Thermal characterization of pure iPS and hybrid gels were performed in a Perkin-Elmer DSC 7 Instrument (Whatmann, Clifton, NJ, USA). Heating and cooling experiments were carried out at a rate of 5 °C·min^{-1}. The test was conducted as follows: Sample was kept at 80 °C during 1 min, then a cooling sweep test was performed up to −5 °C. This temperature was held for 1 min and the sample was subsequently subjected to a heating sweep test going up to 80 °C. The transitions as well as enthalpies generated along the process were determined from the peaks in the thermograms obtained in the heating and cooling modes, respectively.

Rheological studies were performed using a controlled stress oscillatory rheometer TA Instruments AR1000 (New Castle, England). Rough 20 mm diameter parallel plates were used for all the samples: (i) Torque sweep in a range of 0.1 to 20,000 µN·m test was performed to determine the Linear Viscoelastic Region (LVR) at a constant non-destructive frequency of 1 Hz and at −1 °C; (ii) temperature sweep tests in the range of −1 to 70 °C at a heating rate of 2.5 °C·min^{-1} in the LVR were performed.

For the morphological characterization, topographical images were taken using a multimode atomic force microscopy (AFM) with the control system Nanoscope IIe (Veeco Instruments, Santa Barbara, CA, USA) in the tapping mode. For the sample preparation, gels were first dried at room temperature during 24 h prior to their observation. Although some solvent could still be trapped in the polymer, it did not affect the quality of the measurement and obtained images.

3. Results and Discussion

In this work, hybrid gels composed of iPS and [Fe(II) (4-octadecyl-1,2,4-triazole)$_3$(ClO$_4$)$_2$]$_n$ metallo-organic polymer were prepared and the effect of each component concentration in the gel formation was analyzed. For this study, two sets of hybrid gels were prepared: (i) Hybrid gels with a fixed iPS concentration (4%, g/mL) and variable metallo-organic polymer concentration and (ii) hybrid gels with fixed metallo-organic polymer concentration (1.4%, g/mL) and variable iPS content. The fixed concentrations of 4% for iPS and 1.4% for metallo-organic polymer were selected, taking into account that this is the critical concentration above which each polymer is capable to form gel independently. In order to understand the formation and the properties of the obtained hybrid gels, we studied the thermal, viscoelastic, and morphological properties of the hybrid gels.

3.1. Thermal Properties of Hybrid iPS/Metallo-Organic Polymer Gels

Figure 1a shows the thermogram corresponding to pure iPS gel in cis-decaline (4%, g/mL). As observed from the plot, the iPS gel exhibits two endothermic peaks at 18 and 43 °C upon heating and a single exothermic peak at 6 °C when cooling (See Table 1). The temperature at which this first endothermic peak occurs is independent of the iPS concentration, and corresponds to the monotectic transformation derived from the phase separation effect involved in the gelation phenomena [18–21]. The second peak corresponds to the iPS gel melting temperature, which varies with iPS concentration. As it is known, the iPS/cis-decalin is a "non-equilibrium" system. In this type of system, the speed at which a polymer solution is cooled down in a given solvent governs the formation of a determined molecular structure, with a greater or lower degree of organization at the molecular level. Rapid sub-cooling leads to the formation of the gel, while sufficiently slow cooling results in the formation of the crystalline state. Between these two extremes, there are two crystalline phases: Phase s and phase p. Except for the crystalline state, the remaining phases contain intercalated solvent molecules. The gel state has an order of nematic type, while the phase s presents characteristics of a smectic arrangement, in which the folding of polymer chains begins to occur. Phase p is less solvated and can be defined as a peritectic system [18,22,23].

The DSC curve of the pure metallo-organic polymer gel (0.9%, g/mL) in cis-decaline is presented in Figure 1b. The thermogram exhibits an exothermic peak at 11 °C when cooling, which is related to gel formation and an endothermic peak of gel melting at 28 °C upon heating.

In Figure 1c,d, thermograms corresponding to iPS/hybrid gels are shown. Figure 1c corresponds to a representative hybrid gel prepared from a fixed iPS concentration (4%, g/mL) with variable concentration of metallo-organic polymer (in this case, 5.1%, g/mL). As observed from the figure, the hybrid gel exhibits an endothermic peak at 20 °C and a wide endotherm where two peaks are vaguely noticeable upon heating (at approximately 30 and 43 °C). The temperature of the first peak, 20 °C, is similar to that observed in pure iPS gel thermogram (Figure 1a); therefore, this peak corresponds to the monotectic transformation of iPS gel. Regarding the two peaks subtly observed in the wide endotherm, the peak appearing at 30 °C coincides with that observed in Figure 1b corresponding to the melting of the metallo-organic polymer. Similar results are observed in Figure 1d, where the thermograms corresponding to an example of hybrid gel prepared from a fixed metallo-organic polymer concentration (1.4%, g/mL) and variable content of iPS (5.4%, g/mL, for this representative graph) are shown. Three endothermic peaks are depicted in the graph upon heating: 16, 28, and 38 °C (See Table 1), which may correspond to the monotectic transformation of iPS gel, the melting temperature of the metallo-organic polymer gel, and melting temperature of the iPS gel, respectively.

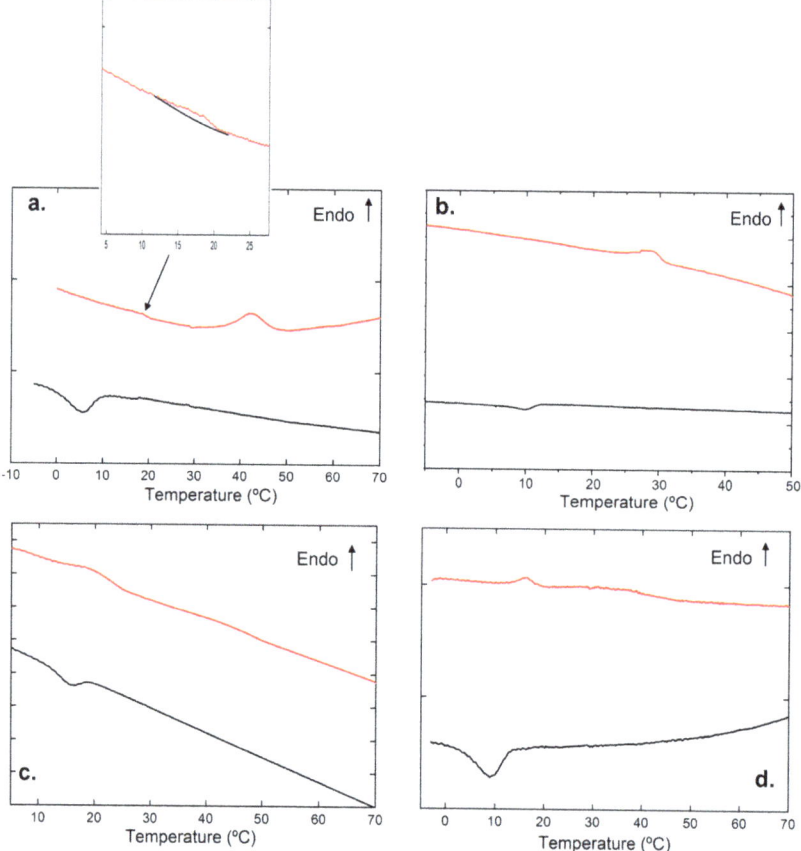

Figure 1. Representative differential scanning calorimetry curves obtained on cooling (red) and on heating (black) for: (**a**) Pure iPS gel in cis-decaline (4%, g/mL) (inset: Amplification of the first endothermic peak at 18 °C), (**b**) metallo-organic polymer gel in cis-decaline (0.9%, g/mL), (**c**) hybrid gel iPS/metallo-organic polymer (iPS = 4%, g/mL and metallo-organic polymer = 5.1%, g/mL), and (**d**) hybrid gel iPS/metallo-organic polymer (metallo-organic polymer = 1.4%, g/mL and iPS = 5.4%, g/mL).

Table 1. Temperatures corresponding to endothermic and exothermic peaks of the pure iPS gel, pure metallo-organic polymer gel, and hybrid gel are collected.

Samples	Endothermic Peaks Temperature (°C)			Exothermic Peak Temperature (°C)
iPS gel	18		43	6
Metallo-organic gel		28		11
Hybrid gel (fixed iPS)	20	30	43	
Hybrid gel (fixed metallo-organic polymer)	16	28	38	

In order to better understand the effect of the composition in the formation of the hybrid gel, in Figure 2a,b, partial phase diagrams are shown. In particular, in Figure 2a, the partial phase diagram of the hybrid gel with a fixed iPS concentration (4%, g/mL) and the pure metallo-organic polymer gel are shown, both as a function of metallo-organic polymer content. As it can be seen, the melting temperature of the pure metallo-organic polymer gel increases slightly with concentration. As previously studied, this is due to an increase in the crosslink density as the concentration of the polymer increases [15,24]. Additionally, it was determined that the physical gel formation was induced by the crystallization of the metallo-organic polymer chains. For the hybrid gel (with a fixed iPS content of 4 g/mL), two different sets of points are represented, each set corresponding to the first and third endothermic peak observed in the thermograms (Figure 1c). The temperature at which the first peak appears remains constant with the increase of the metallo-organic polymer content, indicating that this first peak corresponds to the monotectic transformation of iPS gel.

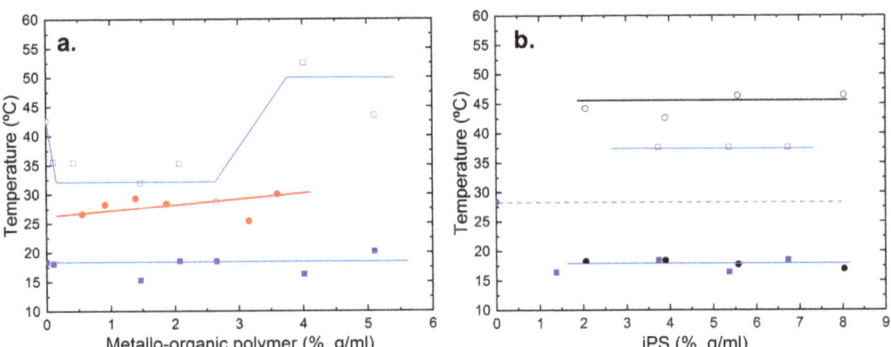

Figure 2. Partial phase diagrams of iPS gels, metallo-organic polymer gels, and hybrid gels obtained upon heating. To build this diagram, first and third endothermic peaks of the hybrid gels corresponding to the iPS gel were chosen: (a) Partial phase diagram obtained upon heating for the iPS/metallo-organic polymer/cis-decaline hybrid gels with a fixed concentration of iPS (4%, g/mL) and variable metallo-organic polymer gel. Pure metallo-organic polymer gel (red circle), first (blue square), and third endothermic peak (blue empty square) of the hybrid gel are plotted. (b) Partial phase diagram obtained on heating for the iPS/metallo-organic polymer/cis-decaline hybrid gels with a fixed concentration of metallo-organic polymer (1.4%, g/mL) and variable concentration of iPS: First (black circle, full symbol) and third endothermic peak of pure iPS (black circle, empty symbol), and first (blue square, full symbol) and third endothermic peak of the hybrid gel (blue square, empty symbol) are represented.

The temperature of the third endothermic peak corresponding to the hybrid gel decreases with respect to the peak related to the pure iPS gel at low metallo-organic polymer concentrations. This result is related to the fact that the metallo-organic polymer is not capable of forming a gel on its own at low concentrations, besides hindering the gelation of iPS. Then, the melting temperature stabilizes for intermediate concentrations of metallo-organic polymer (up to 2.6%, g/mL). At such concentrations, the metallo-organic polymer is capable of forming a gel, however the melting temperatures of the iPS in the hybrid gel are still lower than those of the pure iPS gel. Although the hybrid gel is formed, the metallo-organic polymer forms a "weak" gel at this intermediate concentration, and, therefore, iPS dilution phenomena continue to occur.

When the metallo-organic polymer concentration is high enough to form a gel, and the gelation of the iPS is not hampered, the melting temperature of the iPS in the hybrid gel is higher than the melting temperature of the pure iPS gel. At this point, phase separation arises and separate gelation of the two systems occur.

Similar analysis is performed for the results collected in Figure 2b. In this case, partial phase diagrams as a function of iPS content are shown for pure iPS polymer gel and for hybrid gels with a fixed metallo-organic polymer concentration (1.4%, g/mL). In the case of pure iPS gel, it is observed that both the temperature where the monotectic transition peak appears and the temperature at which the melting peak occurs remain constant with iPS concentration. Regarding the hybrid gel, two series of points corresponding to the first and third peak of the thermograms (Figure 2b) are plotted. From the graph, it is observed that: i) The temperature of the first peak remains constant with the concentration of iPS and ii) the third peak, which also remains constant, only appears above iPS concentrations of 3.5% (g/mL).

As mentioned before, the third peak observed upon heating in the thermograms corresponds to the melting of the iPS gel. The fact that melting temperature of the hybrid gel occurs at temperatures lower than those of iPS gel responds to a dilution phenomenon derived from the presence of the metal-organic polymer in the mixture. However, the temperature at which this peak appears hardly varies with the concentration of iPS, which could indicate the slight influence of the metallo-organic polymer on the gelation of the hybrid system.

Therefore, the obtained results indicate that the hybrid gels are formed by the independent gelation of both systems: iPS/cis-decalin and metallo-organic polymer/cis-decalin. Thus, the limiting parameter for the formation of the hybrid gel is the critical concentration of each polymer independently. Therefore, the hybrid gel will be formed as long as the two polymers are in concentrations above the critical gelation concentrations, which increase in relation to the pure components.

3.2. Viscoelastic Properties of the Hybrid iPS/Metallo-Organic Polymer Gels—Effect of iPS and Metallo-Organic Polymer Concentrations

In Figure 3a,b, the evolution of storage G' and loss modulus G" with the oscillation torque is described for pure iPS gels, pure metallo-organic polymer gels, and for hybrid iPS/metallo-organic polymer gels. The three gels describe a common rheological comportment showing G' higher than G", which is indicative of a solid-like behavior. G' and G" show a plateau up to a critical torque value where both moduli decrease abruptly and crossover modulus (G' = G"). The decreasing and crossover of both moduli entails the transition from a solid-like to liquid-like behavior. From these graphs, we could also determine the linear viscoelastic regime (LVR) where the values of G' and G" are independent of the applied torque [16]. As observed from Figure 3a,b, the values of the storage modulus and the critical torque of the hybrid gels are greater than those of the pure polymers separately. This suggests the interaction or interpenetration of the two networks.

In Figure 4a,b, the evolution of storage and loss modulus with temperature for the pure iPS gel and hybrid iPS/metallo-organic polymer gels is represented. If we focus on the temperature dependence of iPS (Figure 4a,b), two pseudo-equilibrium plateaus can be observed, where G' is independent of the temperature. Between the two plateaus there is a transition, detected by the presence of a maximum in the loss tangent (tan δ) at 18 °C. At higher temperatures, the drop of the modulus and subsequent crossover occurs, which marks the melting of the gel (53 °C). The first transition could correspond to the monotectic transition of the iPS gel, also observed in the thermogram of the pure iPS gel (Figure 1a).

The temperature at which this transition occurs is different depending on the experimental technique used to determine it. Temperatures determined by rheometry and DSC do not coincide, since the speeds at which the heating scans have been carried out are not the same. In DSC, a speed of 5 °C/min has been used, while in rheometry, the speed has been 2.5 °C/min. From the analysis performed on the hybrid gel with a fixed concentration of iPS (4%, g/mL) and variable concentration of metallo-organic polymer (5.1%, g/mL), two plateaus defining two pseudo-equilibrium storage moduli are also observed. In this case, this first transition between both plateaus, also determined by a maximum of the loss tangent, occurs at 25 °C. The crossover of the modulus determining the melting of the hybrid gel occurs at the same temperature as pure iPS gel, 52 °C. Similar conclusions are obtained when comparing pure iPS (4%, g/mL) with the hybrid gel with a fixed concentration of metallo-organic polymer (1.4%, g/mL) and variable concentration of iPS (3.7%, g/mL) (Figure 4b). A difference is only found in the temperature at which the melting takes place: In the hybrid gel, melting occurs at 33 °C, being significantly lower than the melting temperature of the pure iPS gel (53 °C).

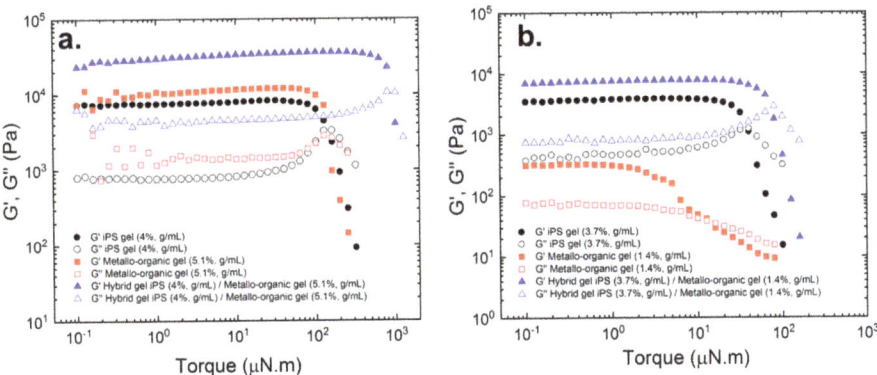

Figure 3. Storage (full symbols) and loss (empty symbols) modulus as a function of oscillation torque, at a constant frequency of 1 Hz and a temperature of −1 °C for: (**a**) Pure iPS gel, 4%, g/mL (black), pure metallo-organic polymer gel 5.1%, g/mL (red), and representative hybrid gel (blue) with a fixed concentration of iPS (4%, g/mL) and variable concentration of metallo-organic polymer (5.1%, g/mL). (**b**) Pure iPS gel, 3.7% g/mL (black), pure metallo-organic polymer gel, 1.4% g/mL (red), and representative hybrid gel (blue) with a fixed concentration of metallo-organic polymer (1.4% g/mL) and variable concentration of iPS (3.7%, g/mL).

In Figure 5a, we have represented the equilibrium storage modulus G' as a function of metallo-organic polymer concentration for hybrid gels with fixed iPS concentration (4%, g/mL) and for pure metallo-organic polymer gels. As it can be observed, the equilibrium storage modulus of pure metal-organic polymer gels increase with concentration; besides, the value of storage modulus appears to be above the equilibrium storage modulus of the pure iPS gel of 4% (g/mL) at a concentration of metal-organic polymer close to 3.6% (g/mL). In the case of the hybrid gel (with fixed iPS concentration), the equilibrium storage modulus decreases up to a concentration of 0.5% (g/mL), from which it begins to increase, ending at a value that is higher than the one shown by pure iPS gel (4%, g/mL) above a metallo-organic polymer concentrations of 1% (g/mL). The equilibrium storage modulus of the hybrid gels is always higher than those of the pure metal-organic polymer gels with the same metallo-organic polymer concentration. However, the equilibrium storage modulus of the hybrid gels with metallo-organic polymer concentrations below 1% (g/mL) are lower than the equilibrium value of the pure iPS gel of 4% (g/mL). At such low concentrations (>1%, g/mL), the metallo-organic polymer is not able to form gel on its own, furthermore preventing the gelation of the iPS. This could be explained by a hindrance of the rigidification of the iPS chains or a process of dilution of the iPS in the mixture because of the presence of metallo-organic polymer. When the concentration of metallo-organic polymer in the hybrid gel is enough for the its gelation to take place, then the equilibrium storage modulus increases. This augment could be due to the formation of an interpenetrated network of iPS and metallo-organic polymer gels.

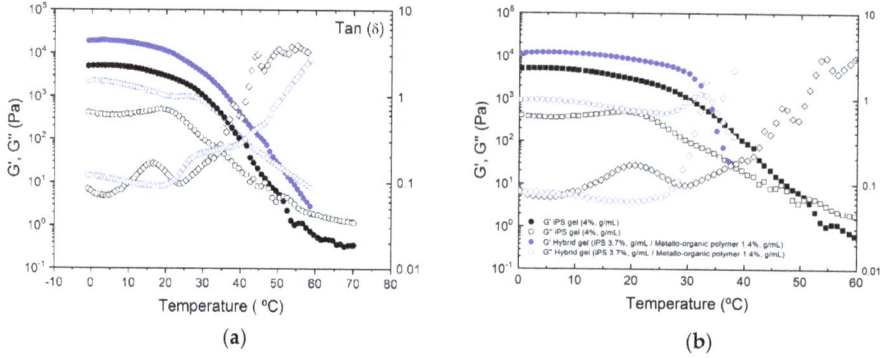

Figure 4. Storage (full symbols) and loss (empty symbols) modulus as a function of oscillation torque measured for: (**a**) Pure iPS gel, 4%, g/mL (black), hybrid gel (blue) with a fixed concentration of iPS (4%, g/mL) and variable concentration of metallo-organic polymer (5.1%, g/mL) and (**b**) pure iPS gel, 4%, g/mL (black), and representative hybrid gel (blue) with a fixed concentration of metallo-organic polymer (1.4%, g/mL) and variable concentration of iPS (4%, g/mL).

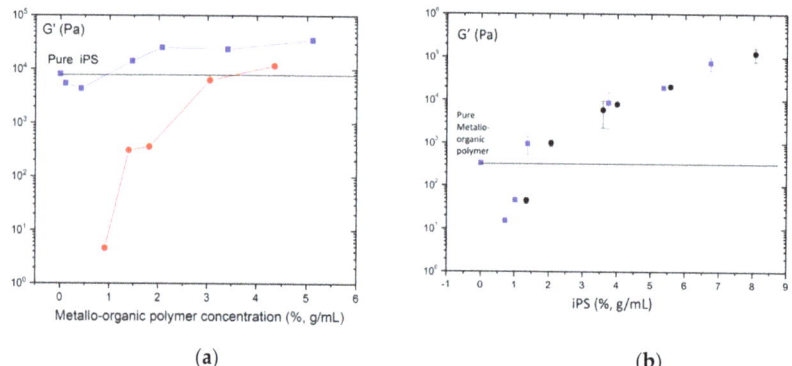

Figure 5. (**a**) Equilibrium storages modulus plateau as a function of metallo-organic polymer concentration for hybrid gels with a fixed iPS concentration of 4%, g/mL (blue square) and for pure metallo-organic polymer gels (red circle); (**b**) equilibrium storages modulus plateau as a function of iPS concentration for hybrid gels with a fixed metallo-organic polymer concentration of 1.4%, g/mL (blue squares) and for pure iPS gels (black circle).

We also analyzed the effect of iPS concentration in the equilibrium storage modulus of hybrid iPS/metallo-organic polymer gel with fixed metallo-organic polymer content (1.4%, g/mL). As observed from Figure 5b, the equilibrium storage modulus of pure iPS gels increases with the concentration of iPS. In the case of the hybrid gels, those with an iPS concentration below 1.2% (g/mL) have lower equilibrium storage modulus values than the pure metal-organic polymer gel. At iPS concentrations below 1.2% (g/mL), the iPS is not capable of gelifying and it could probably inhibit the gelation of the metallo-organic polymer. However, above 1.2% iPS concentration, hybrid gels present equilibrium storage modulus higher than the storage modulus of the pure metallo-organic polymer gel. Moreover, above iPS concentration of 3.5%, g/mL, the equilibrium storage modulus of both pure iPS gel and hybrid gel become similar. This could be due to the fact that the value of the equilibrium storage modulus of the pure metal-organic polymer gel is too low to contribute significantly to the value of the equilibrium modulus of the hybrid gel.

In Figure 6, the double logarithmic plots of the equilibrium storages modulus as a function of metallo-organic polymer concentration (Figure 6a) and iPS concentration (Figure 6b) are represented for both hybrid gels and pure metallo-organic polymer and iPS gels, respectively. The obtained plots have been analyzed using a theoretical model based on de Gennes scaling law, in which the moduli are related to the concentration as follows: $G \propto C^n$ [25]; n being an exponent that depends on the conformation of the polymer chains linking the connection points.

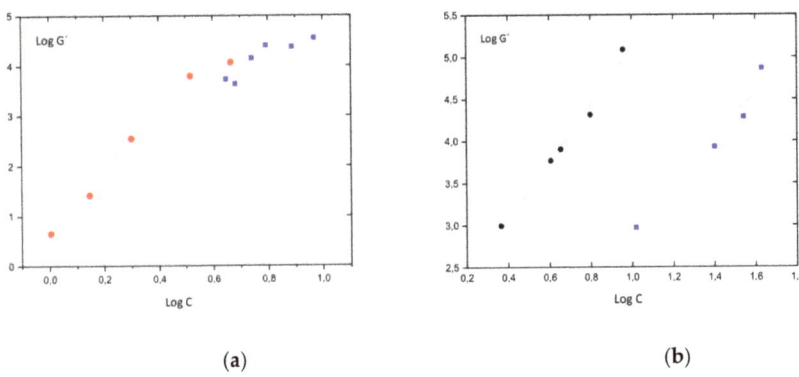

(a)   (b)

Figure 6. (a) Double logarithmic plot of the equilibrium storages modulus plateau as a function of metallo-organic polymer concentration for hybrid gels with a fixed iPS concentration of 4%, g/mL (blue square) and for pure metallo-organic polymer gels (red circle) and (b) as a function of iPS concentration for hybrid gels with a fixed metallo-organic polymer concentration of 1.4%, g/mL (blue squares) and for pure iPS gels (black circle).

As it has already been indicated in our previous works, these correlations were developed for chemical gels but they were already applied for physical gels [15]. Since the storage modulus of the studied gels describe two pseudo-equilibrium moduli where $G \neq G(T)$, it is possible to indicate that the obtained gels are formed by networks presenting enthalpic elasticity. For this type of network, there is an approximation that correlates the equilibrium storage modulus with the concentration through the fractal dimension (υ-1) of the objects linking the connection points: $G \propto C^{(3\upsilon + 1)/(3\upsilon - 1)}$ [26]. We have applied this approximation to pure metallo-organic polymer (Figure 6a) and pure iPS gels (Figure 6b) as well as to the hybrid gels with fixed iPS polymer concentration 4%, g/mL (Figure 6a) and fixed metallo-organic polymer concentration 1.4%, g/mL (Figure 6b) and obtained the fractal dimensions shown in Table 2.

Table 2. Fractal dimension obtained by the correlation $G \propto C^{(3\upsilon + 1)/(3\upsilon - 1)}$ extracted from Figure 6.

Fractal Dimension	Pure iPS Gel	Pure MOP * Gel	Hybrid Gels	
			Fixed iPS Concentration (4%, g/mL)	Fixed MOP * Concentration (1.4 g/mL)
υ^{-1}	1.5	2.5	1.4	1.5

* MOP: Metallo-organic polymer.

In the case of pure iPS gel, the obtained value is 1.5, which would correspond to polymer fibers swollen by the solvent. If we focus on the fractal dimensions of the hybrid gels, the obtained values (1.4 for hybrid gels with fixed iPS concentration and 1.5 for hybrid gels with fixed metallo-organic polymer concentration) are similar to the fractal dimension of pure iPS gel, and far from the fractal dimension of the pure metallo-organic polymer gel (2.5) [15]. This would indicate that the iPS determines the structure and the viscoelastic properties of the hybrid gel, even though the two networks are formed independently.

When a material is deformed, its elastic behavior can be outlined simply by the Hook's law. However, if the material is multicomponent, as it is the case for the hybrid gel that is composed of two polymeric gels (iPS and metallo-organic polymer), there is more than one force constant responsible for the final module of the system. Therefore, it is necessary to find a model that could consider these constants so that the resulting force constant describes the elastic behavior of the hybrid gel. Therefore, in order to determine the contribution that each individual gel (iPS gel and metallo-organic polymer gel) has in the storage modulus of the hybrid gel, we have suggested the following two theoretical models. (i) Sum of the individual gel modulus in series: The force constants will act independently against an external force, initially deforming the less rigid spring, and later, the one with the highest rigidity (see the corresponding scheme in Figure 7). (ii) Sum of the modules of the individual gels in parallel: The force constants of the individual gels are interdependent, in such a way that the elastic response of the material is limited by the system whose force constant is higher (see the corresponding scheme in Figure 7).

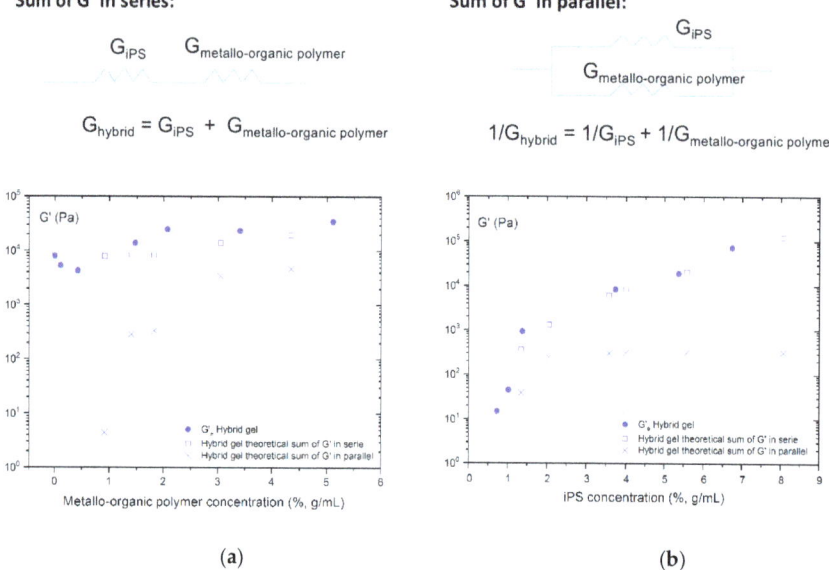

Figure 7. Pictorial scheme of the applied models. Representation of the experimental equilibrium storage modulus as a function of (**a**) metallo-organic polymer concentration (●) and (**b**) iPS concentration (●) and their corresponding theoretical sum of modules in parallel (x) and theoretical sum of the modules in series (□).

In Figure 7a,b, experimentally obtained equilibrium storage moduli of the hybrid gel are compared with the values determined through the application the two theoretical models from the experimental data. As concluded from both graphs, the theoretical model that best describes the experimental elastic behavior of the hybrid gel is the one that considers the elastic behavior of the hybrid gel as the sum of individual gel's storage modulus in series. That is, the hybrid gel would consist of two networks of pure polymers independent of each other.

For the sake of comparison, in Figure 8a, we have collected the melting temperatures (T_m) obtained for pure metallo-organic polymer gels and iPS/metallo-organic polymer hybrid gels with a fixed iPS concentration of 4%, g/mL and represented as a function of metallo-organic polymer concentration

(%, g/mL). From the figure, we observe how the melting temperature values of the metallo-organic polymer gels increase with the concentration in the studied range of concentrations. In contrast, the melting temperatures of the hybrid gels decreases up to a concentration of metallo-organic polymer of 0.4%, g/mL; beyond this concentration, the melting temperatures increase to values slightly above the pure iPS gel of 4% (g/mL). The melting temperatures of the hybrid gels are lower than those of the pure iPS gel (4%, g/mL) in almost the entire range of metallo-organic polymer concentrations. Therefore, at low metallo-organic polymer concentrations, dilution phenomena may occur, preventing the gelation of the iPS, which causes the melting temperature of the hybrid gel to decrease. This confirms the idea of the formation of the hybrid gels, from the gelation of both iPS and metallo-organic polymers independently, as long as the two polymers are in concentrations above the critical gelation concentrations.

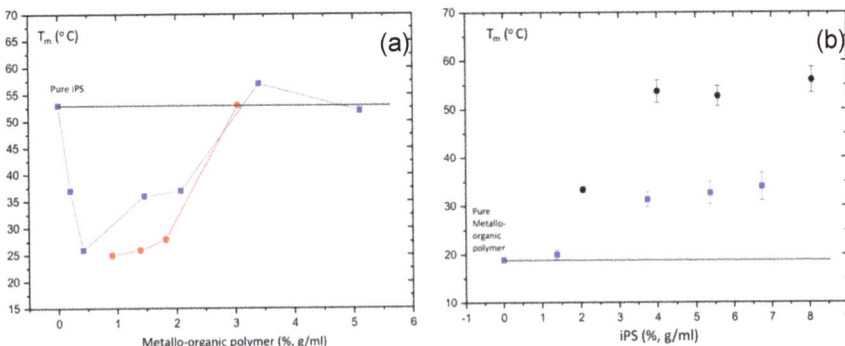

Figure 8. (**a**) Melting temperature (Tm) as a function of metallo-organic polymer concentration (%, g/mL) for pure metallo-organic polymer gel (red circle) and iPS/metallo-organic polymer hybrid gel with a fixed iPS concentration of 4%, g/mL (blue squares). (**b**) Melting temperature (Tm) as a function of iPS concentration (%, g/mL) for pure iPS polymer gel (black circle) and iPS/metallo-organic polymer hybrid gel with a fixed metallo-organic polymer concentration of 1.4%, g/mL (blue squares).

Figure 8b shows the melting temperatures of pure iPS gels and hybrid iPS/metallo-organic polymer gel with a fixed metallo-organic polymer concentration (1.4%, g/mL) as a function of iPS concentration. As observed, melting temperature values of pure iPS gels increase with the concentration of iPS. This increase seems to stabilize at concentrations of 4%, g/mL, above which Tm keeps constant. Similar to what occurs for pure iPS gels, the melting temperatures of the hybrid gels also increase with iPS concentration until it reaches a concentration (close to 4%, g/mL) at which it remains constant. Besides, hybrid gels present melting temperatures higher than the pure metallo-organic polymer gel of 1.4%, (g/mL) (18 °C) but significantly lower than the melting temperature values of the pure iPS gels at each iPS concentration.

When the concentration of iPS in the hybrid gel is not enough for the polymer to undergo gelation on its own, what we are really observing is the melting of the metallo-organic polymer. But, with enough concentration of iPS to gelate on its own, the melting temperature trend observed for the hybrid gels is the same as that of the pure iPS gels. The main difference is that the melting temperature values of hybrid gels are much lower than those of pure iPS gels. It is probably that the observed melting may correspond to the iPS gel, but the temperatures at which this melting occurs are lower due to a dilution effects in the blend.

3.3. Morphology of the Hybrid iPS/Metallo-Organic Polymer Gels

We performed a morphological study by means of AFM technique and obtained micrographs corresponding to the pure iPS, pure metallo-organic polymer, and hybrid iPS/metallo-organic polymer xerogels, as shown in Figure 9. If we focus on the images corresponding to the pure iPS and pure metallo-organic polymer xerogels, in both micrographs, a network morphology typical of a gel-like structure can be observed. The iPS xerogel appears to consist of fibers with an average diameter of 25 ± 5 nm (Figure 9a). These values are in accordance with those observed in the literature [27]. In the case of metallo-organic polymer xerogel, this is formed by fibers with a mean diameter of 10.3 ± 2.5 nm (Figure 9b).

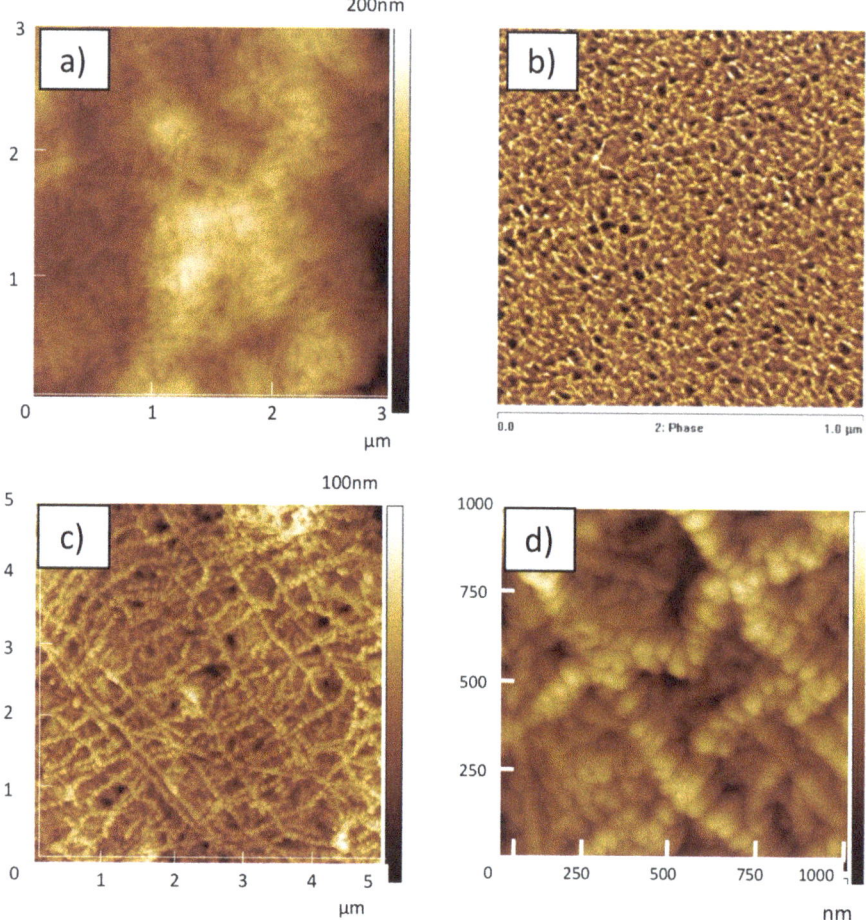

Figure 9. Representative atomic force microscopy (AFM) micrographs corresponding to (**a**) pure iPS xerogel, (**b**) pure metallo-organic polymer xerogel, and (**c**,**d**) hybrid iPS/metallo-organic polymer gel (with iPS/metallo-organic polymer ratio of 97/3).

Regarding micrographs corresponding to hybrid iPS/metallo-organic polymer xerogel (Figure 9c,d), two networks formed with fibers of different sizes are observed. We have measured the diameters of the fibers found on each of the networks and determined that fibers of one network present a diameter of 26 ± 12 nm. This fiber diameter size is similar to the size of iPS xerogel fibers, therefore we could consider that this network could correspond to the one formed by iPS xerogel. The other observed network is formed by fibers with an average diameter of 105 ± 23 nm. Besides, the fibers forming this network seems to be polymer agglomerates, probably agglomerates of the metallo-organic polymer xerogel fibers. Therefore, from the morphological observation, we could consider that the gels of the two polymers grow independently of each other in the hybrid gel.

4. Conclusions

By mixing homogeneous solutions of metallo-organic polymer and iPS in cis-decalin separately, stirring at 100 °C, and cooling said solutions to 0 °C, hybrid gels of metallo-organic polymer and iPS in cis-decaline can be prepared. The obtained hybrid gels can be melted above a certain temperature and re-formed upon cooling below a certain temperature (gelling temperature) at a sufficient cooling rate. These are, therefore, thermo-reversible hybrid gels.

From the study of the thermal properties of the hybrid gels carried out by DSC, three transitions have been determined upon heating: The monotectic transition of the iPS gel, the melting of the iPS gel, and the melting of the metallo-organic polymer gel. The latter two transitions are not always clearly observed, since the endotherms are very wide. All seem to indicate that the gels of the two polymers are formed independently, without affecting each other much, as long as the two polymers are in concentrations above the corresponding critical gelation concentrations.

Through the study of the viscoelastic properties of the hybrid gels, an analogy has been observed between the viscoelastic behavior of hybrid gels and pure iPS gels. Both torque sweeps and temperature sweeps have similar shapes for the different systems. Specifically, in the temperature scans it is possible to observe the two iPS gel transitions seen in the thermal measurements: Monotectic transition and gel melting. Both the equilibrium storage modulus and the melting temperature of the hybrid gels are greater than the equilibrium modulus and the melting temperatures of the gels of the pure polymers when the polymer concentrations are high enough to avoid dilution phenomena in the mixtures. This would indicate the formation of an interpenetrated network of polymer gels, which grow independently.

To determine the structure of the hybrid gels, the approximation for systems with enthalpic elasticity of the De Gennes model was used. By this method, a fractal dimension of 1.4 has been determined for hybrid gels with constant concentration of iPS. This value is similar to that obtained for pure iPS gels and is well below that obtained for pure polymer-metal-organic gels. For hybrid gels with a constant concentration of metallo-organic polymer, the obtained fractal dimension is 1.5, similar to that of the pure iPS gel. In both cases, it can be said that the iPS is the polymer that determines the final structure of the hybrid gel.

Attempts have been made to determine the contributions of the gels of the pure polymers separately to the storage module (G') of the hybrid gels. It has been observed how the elastic behavior of the hybrid gels fitted best to a model in which the elastic moduli of the gels are added in series. This behavior is typical of systems with two independent phases, in which the iPS governs the elastic behavior.

Through AFM, the presence of two independent networks has been observed in the hybrid xerogel. These networks correspond to the networks of the gels of the two separate polymers, which are growing independently in the hybrid gel forming two phases. These hybrid gels, therefore, are interpenetrated organic/metallo-organic networks.

Author Contributions: Conceptualization, D.L.; methodology, M.R. and D.L.; investigation, D.L, M.L. and C.E.; writing—original draft preparation, C.E. and D.L.; writing—review and editing, C.E. and D.L.; supervision, D.L.

Funding: This research received no external funding.

Acknowledgments: C.E. acknowledges IJCI-2015-26432 contract from MINEICO.

Conflicts of Interest: The authors declare no conflict of interest.

References

1. Lahav, M.; van der Boom, M.E. Polypyridyl metallo-organic assemblies for electrochromic applications. *Adv. Mater.* **2018**, *30*, 1706641. [CrossRef]
2. Dov, N.E.; Shankar, S.; Cohen, D.; Bendikov, T.; Rechav, K.; Shimon, L.J.W.; Lahav, M.; Van Der Boom, M.E. Electrochromic metallo-organic nanoscale films: Fabrication, color range, and devices. *J. Am. Chem. Soc.* **2017**, *139*, 11471–11481.
3. Kahn, O. Spin-transition polymers: From molecular materials toward memory devices. *Science* **1998**, *279*, 44–48. [CrossRef]
4. Southon, P.D.; Liu, L.; Fellows, E.A.; Price, D.J.; Halder, G.J.; Chapman, K.W.; Moubaraki, B.; Murray, K.S.; Letard, J.F.; Kepert, C.J. Dynamic interplay between spin-crossover and host-guest function in a nanoporous metal-organic framework material. *J. Am. Chem. Soc.* **2009**, *131*, 10998–11009. [CrossRef]
5. Gamez, P.; Costa, J.S.; Quesada, M.; Aromi, G. Iron spin-crossover compounds: From fundamental studies to practical applications. *Dalton Trans.* **2009**, *38*, 7845–7853. [CrossRef]
6. Boukheddaden, K.; Ritti, M.H.; Bouchez, G.; Sy, M.; Dîrtu, M.M.; Parlier, M.; Linares, J.; Garcia, Y. Quantitative contact pressure sensor based on spin crossover mechanism for civil security applications. *J. Phys. Chem. C* **2018**, *122*, 7597–7604. [CrossRef]
7. Kumar, K.S.; Ruben, M. Emerging trends in spin crossover (sco) based functional materials and devices. *Coord. Chem. Rev.* **2017**, *346*, 176–205. [CrossRef]
8. Halcrow, M.A. *Spin-Crossover Materials. Properties and Applications*; Halcrow, M.A., Ed.; John Wiley & Sons Ltd.: London, UK, 2013.
9. Kumar, K.S.; Studniarek, M.; Heinrich, B.; Arabski, J.; Schmerber, G.; Bowen, M.; Boukari, S.; Beaurepaire, E.; Dreiser, J.; Ruben, M. Engineering on-surface spin crossover: Spin-state switching in a self-assembled film of vacuum-sublimable functional molecule. *Adv. Mater.* **2018**, *30*, 1705416. [CrossRef]
10. Shalabaeva, V.; Mikolasek, M.; Manrique-Juarez, M.D.; Bas, A.-C.; Rat, S.; Salmon, L.; Nicolazzi, W.; Molnár, G.; Bousseksou, A. Unprecedented size effect on the phase stability of molecular thin films displaying a spin transition. *J. Phys. Chem. C* **2017**, *121*, 25617–25621. [CrossRef]
11. Gaspar, A.B.; Seredyuk, M. Spin crossover in soft matter. *Coord. Chem. Rev.* **2014**, *268*, 41–58. [CrossRef]
12. Voisin, H.; Aimé, C.; Vallée, A.; Bleuzen, A.; Schmutz, M.; Mosser, G.; Coradin, T.; Roux, C. Preserving the spin transition properties of iron-triazole coordination polymers within silica-based nanocomposites. *J. Mater. Chem. C* **2017**, *5*, 11542–11550. [CrossRef]
13. Echeverría, C.; Rubio, M.; Mitchell, G.R.; López, D. Structure of a spin-crossover Fe(ii)-1,2,4-triazole polymer complex gel in toluene. Small angle neutron scattering and viscoelastic studies. *Eur. Polym. J.* **2014**, *53*, 238–245. [CrossRef]
14. Echeverria, C.; Rubio, M.; Mitchell, G.R.; Roig, A.; López, D. Hybrid polystyrene based electrospun fibers with spin-crossover properties. *J. Polym. Sci. Part B Polym. Phys.* **2015**, *53*, 814–821. [CrossRef]
15. Rubio, M.; López, D. Effect of solvent on the gelation properties of a metallo-organic polymer of [Fe(ii) (4-octadecyl-1,2,4-triazole)$_3$(ClO$_4$)$_2$]$_n$. *Eur. Polym. J.* **2009**, *45*, 3339–3346. [CrossRef]
16. Rubio, M.; Hernández, R.; Nogales, A.; Roig, A.; López, D. Structure of a spin-crossover Fe(ii)–1,2,4-triazole polymer complex dispersed in an isotactic polystyrene matrix. *Eur. Polym. J.* **2011**, *47*, 52–60. [CrossRef]
17. Dasgupta, D.; Srinivasan, S.; Rochas, C.; Ajayaghosh, A.; Guenet, J.M. Hybrid thermoreversible gels from covalent polymers and organogels. *Langmuir* **2009**, *25*, 8593–8598. [CrossRef] [PubMed]
18. Guenet, J.-M. Contributions of phase diagrams to the understanding of organized polymer-solvent systems. *Thermochim. Acta* **1996**, *284*, 67–83. [CrossRef]
19. He, X.; Herz, J.; Guenet, J.M. The physical gelation of a multiblock copolymer: Effect of solvent type. *Macromolecules* **1989**, *22*, 1390–1397. [CrossRef]
20. Reisman, A. *Phase Equilibria: Basic Principles, Applications, Experimental Techniques/Arnold Reisman*; Academic Press: New York, NY, USA, 1970.
21. Guenet, J.M.; McKenna, G.B. Thermoreversible gelation of isotactic polystyrene: Thermodynamics and phase diagrams. *Macromolecules* **1988**, *21*, 1752–1756. [CrossRef]
22. Guenet, J.M.; Menelle, A.; Schaffhauser, V.; Terech, P.; Thierry, A. Isotactic polystyrene/cis-decaline mixtures: Phase diagram and molecular structures. *Colloid Polym. Sci.* **1994**, *272*, 36–47. [CrossRef]

23. Klein, M.; Mathis, A.; Menelle, A.; Guenet, J.M. Structures in isotactic polystyrene solutions near the physical gelation threshold. *Macromolecules* **1990**, *23*, 4591–4596. [CrossRef]
24. Guenet, J.M. *Thermoreversible Gelation of Polymers and Biopolymers*; Academic Press: London, UK, 1992; p. 287.
25. Gennes, P.G.D. *Scaling Concepts in Polymer Physics*; Cornell University Press: London, UK, 1979.
26. Jones, J.L.; Marques, C.M. Rigid polymer network models. *J. Phys. Fr.* **1990**, *51*, 1113–1127. [CrossRef]
27. Guenet, J.M.; Wittmann, J.C.; Lotz, B. Thermodynamic aspects and morphology of physical gels from isotactic polystyrene. *Macromolecules* **1985**, *18*, 420–427. [CrossRef]

© 2019 by the authors. Licensee MDPI, Basel, Switzerland. This article is an open access article distributed under the terms and conditions of the Creative Commons Attribution (CC BY) license (http://creativecommons.org/licenses/by/4.0/).

Article

Influence of Incorporated Polydimethylsiloxane on Properties of PA66 Fiber and Its Fabric Performance

Wei Peng, Ying Qian, Tong Zhou, Shenglin Yang, Junhong Jin and Guang Li *

State Key Laboratory for Modification of Chemical Fibers and Polymer Materials, College of Materials Science and Engineering, Donghua University, Shanghai 201620, China; pwpz321654@hotmail.com (W.P.); viskyying@163.com (Y.Q.); 15921876270@163.com (T.Z.); slyang@dhu.edu.cn (S.Y.); jhkin@dhu.edu.cn (J.J.)
* Correspondence: lig@dhu.edu.cn

Received: 18 September 2019; Accepted: 16 October 2019; Published: 23 October 2019

Abstract: Poly(hexamethyllene adipamide), PA66 fiber has played an important role in varied industrial applications, and its corresponding product would become more competitive if some extra value was added to PA66 fiber. In this article, polydimethylsiloxane (PDMS) was used as an additive to prepare PA66/PDMS blend fibers through melt blend spinning carried out by a screw extruder spinning machine. When the amount of incorporated PDMS was 0.5–3 wt %, the blend melt demonstrated good spinning ability, and the PA66/PDMS blend fibers exhibited excellent mechanical property and reduced hot shrinkage. Moreover, the crystallization and melting behavior of PA66 in the blend fibers turned out to be not affected by the existence of PDMS. In addition, the contact angle of water on the blend fiber surface became larger, while the value of friction coefficient on the surface of fibers got lower with increasing PDMS content in the blend fibers. After evaluating the fabric woven by PA66/PDMS blend fibers using the KES-F KES-FB-2 fabric measuring system, it was found that as PDMS content increased, the flexural rigidity and bending hysteresis would be lower, yet elasticity rate of compression work would be higher, which explained how the fabric composed of the blend fiber performed better in terms of softness and elasticity.

Keywords: poly(hexamethyllene adipamide); polydimethylsiloxane; blending fiber; fabric performance

---

1. Introduction

Polydimethylsiloxane (PDMS), commonly known as silicone, has been generally used as an additive in polymer processing [1,2]. PDMS can improve flow-ability of melt which is especially required under some circumstances such as processing some products with thin walls and complex shapes [3], where full filling of injected melt is needed. PDMS can also reduce friction between products and equipment [4], keeping the surface of products bright and clean to make product quality high and more competitive.

Hager et al. [5] prepared a multifunctional additive based on PDMS of ultrahigh molecular weight. He found this additive highly effective regarding mobility of thermoplastic processing, especially applicable to polyolefin processing. Besides, this additive not only strengthened impact and tensile strength of products, but made the surface highly polished and resistant to abrasion as well. Jin et al. [6] prepared a long branched PDMS grafted on PE chains (PDMS-g-PE) and used it as plasticizer to improve the melt processing of HDPE high-density polyethylene (HDPE). It was discovered that melt flow index would decrease when PDMS-g-PE content was 2 wt %, suggesting melt mobility of HDPE was improved to a certain extent. In practical use, PDMS has also been commonly used to coat fibers and fabrics, acting as an oil or finishing agent to make fiber and its product have a low friction coefficient and a smooth feeling [7–9]. At the same time, it was noticed that PDMS on the surface can be easily washed away during the dyeing and washing process, which would not only invalidate the function of PDMS itself, but also bring in negative consequences to the environment.

The purpose of this study is to incorporate PDMS into PA66 during the melt-blend spinning process through a screw extrude machine. Thus, PDMS can be distributed both inside and on the surface of PA66 fiber, leading to low friction on the fiber surface. Moreover, PDMS will move onto the surface gradually in practical use, finally realizing the enrichment of PDMS on fiber surface. As a result, the surface friction coefficient could be reduced and the water contact angle on fiber surface could be enlarged, in addition, the fibrous fabric would become soft and fluffy with function of waterproofing and antifouling as well due to the presence of PDMS.

2. Experimental

2.1. Materials

PA66 chips were from Shenma Group, Pindingshan, China, FYR-27, PDMS master-batch, MB50011, from Dow Corning Ltd. (Soochow, China).

2.2. Preparation of PA66/PDMS Blend Fibers

Blend chips of PA66 with addition of PDMS (0.5–3.0 wt %) were dried in vacuum at 110 °C for 48 h. Then, PA66/PDMS blend fibers were obtained by melt-blend spinning using ABE-25 spinning machine. Each as-spun fiber was received at 1000 m/min of take-up speed, and then stretched out using through TF-100 stretching equipment with heat board at 60 °C and heat plate at 120 °C, with drawing ratio of 3.5, 3.7, 3.9, and 4.1, respectively.

2.3. Methods

2.3.1. Capillary Rheological Performance Measurement of PA66/PDMS Blends

First, PA66/PDMS blend chips were obtained through melt-mixing by twin-screw extruder where PA66 and PDMS were dried at 110 °C for 48 h to remove water in advance. The capillary rheological behavior was investigated using Marrin RH2000 Capillary Rheometer at 275, 280, and 285 °C, respectively, with a capillary radius of 0.5 mm, L/D ratio 16:1, and sheer rate in the range of 100–10,000 s^{-1}.

2.3.2. Differential Scanning Calorimetry (DSC)

DSC spectra were recorded on TA-Q20 DSC apparatus (city, country Shanghai, China). In the test, 5–10 mg dried samples were first heated to 300 °C, kept for 3 min at that temperature, then cooled down to room temperature at a rate of 10 °C/min. Finally, the samples were heated again to 300 °C at the same rate of 10 °C/min.

2.3.3. Thermal Shrinkage

Thermal shrinkage of PA66/PDMS blend fibers was measured manually according to GB/T6505-2001. Before testing, fibers were balanced at standard environment for 3 h; next they were treated in an oven at 190 °C for 15 min, and then balanced for 4 h. Every sample was examined three times to averagely calculate thermal shrinkage as the formula: $S = (L_0 - L_s)/L_0 \times 100\%$, where S is thermal shrinkage (%), L_0 is the original length of fiber (mm), L_s is the length after heating treatment (mm).

2.3.4. Preparation of PA66/PDMS Blend Fibrous Fabric

PA66/PDMS blend fibers containing different ratio of PDMS with the same drawing ratio of 3.5 were woven into simple plain texture fabric, respectively, in handed shuttle loom in our lab. Each fabric was woven at the same condition by manual control as best as one can.

2.3.5. Silicon Analysis by Energy Dispersive Spectra-Meter (EDS)

PA66/PDMS blend fibers were washed by acetone and distilled water to remove oil on the surface, then dried and kept under constant temperature of 20 °C and humidity of 65% prior to test. With help from IncaX-Max EDS (Hitachi G, Japan), the presence of PDMS (Si) and its distribution was investigated on cross section and the surface of fibers.

2.3.6. Water Contact Angle on Fiber

The German OCA40Micro optical contact angle measuring device (Germany) was used. Before the experiment, the blend fiber was washed by acetone and distilled water for surface degreasing, and then drying and balancing under temperature of 20 °C and relative humidity of 65%. During testing, the volume of water drop was controlled at 300 µL and 0.65 Pa. The average value was taken from 30 test results for each sample.

2.3.7. Friction Coefficient of Fiber Surface

XCF-1A fiber friction meter (mode: friction roller rotation) was employed for measurement of friction coefficient of fiber surface. The blend fiber was washed by acetone and distilled water, then dried and kept under constant temperature of 20 °C and humidity of 65% prior to the test. The pre-tension was set as 0.1 cN, rotation speed 30 rpm, and decline rate of friction roller 10 mm/min. The average value was taken based on 15 test runnings for each sample.

2.3.8. Bending Properties of Fabrics

According to GB/T 18318.5-2009, KES-FB-2 equipment was used to examine bending properties of fabrics, from which flexural rigidity (B) and bending hysteresis (2HB) could be obtained for further analysis. Before the test, the standard temperature and humidity conditioning were implemented in accordance with GB/T 6529.

2.3.9. Compression Performance of Fabrics

As instructions specified in GB/T 24442.2-2009 (part 2), compression performance was tested by KES-FB-3 equipment (city, countryShanghai, China) to obtain compression work (WC) and elasticity rate of compression work (RC). Temperature and humidity conditioning were implemented in accordance with GB/T 6529 before the test.

3. Results and Discussion

3.1. Effects of PDMS on Flowing Behavior of PA66 Melt

High melt flow ability was preferred during the fiber fabrication. When PDMS was added into PA66, the blend melt behavior was investigated using capillary rheometer. As illustrated in Figure 1a, the fluid curves of pristine PA66 and three PA66/PDMS mixed melts with addition of 1, 2, and 3 wt % PDMS are present. Similar to PA66, the apparent viscosity of mixed melts falls as shear rate increases; besides, the decreased extent is positively correlated with the content of PDMS added, suggesting a positive effect on flow ability of PDMS addition. This may be attributed to the mutual repulsion between Si–O and methyl groups in PDMS chains. As shown in Figure 1b, the outward arrangement of methyl groups, acting as a screen, can result in low surface energy and low surface tension. Thus, it can be seen as a lubricator reducing the force and entanglement density among PA66 molecular chains, and further reducing the apparent viscosity of the system. Therefore, the addition of certain amounts of PDMS to PA66 is able to improve melt flow ability and help the formation of PA66 fiber.

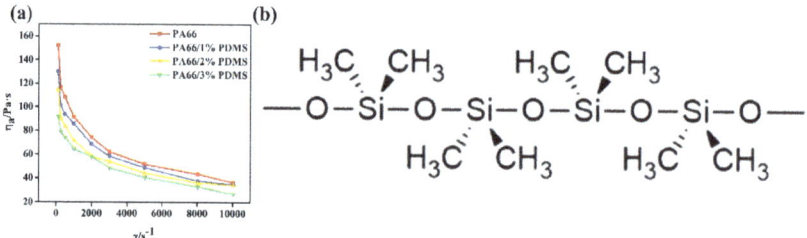

Figure 1. The flow curves of PA66 and PA66/PDMS melts (a), and chemical conformation of PDMS (b).

3.2. Thermal Properties of PA66/PDMS

Figure 2 exhibits the crystallization and melting curves of the mixed PA66/PDMS melt with different PDMS contents, from which no significant effect is observed on crystallization and melting of PA66 coming from PDMS addition. Crystallization temperature of PA66 stabilized at 232–233 °C, and the melting point of PA66 stayed at 262–263 °C. The existence of PDMS is able to enhance mobility of the melt, and at the same time gives it good thermostability and does not influence crystallization and fusion of PA66. This can be beneficial to formation and application of modified PA66 fibers.

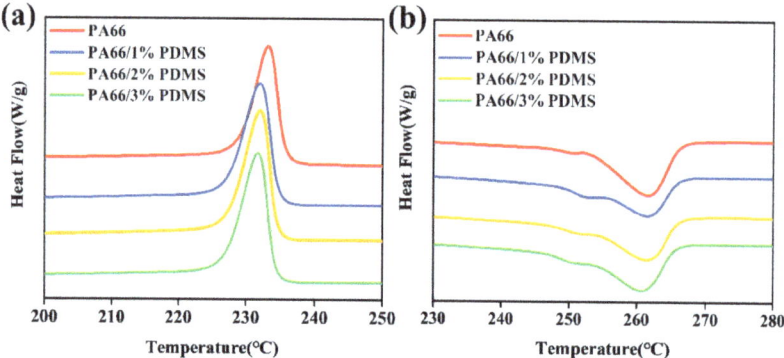

Figure 2. Crystallization (a) and melting (b) of PA66/PDMS blends.

3.3. Mechanical Properties of PA66/PDMS Blend Fibers

The PA66/PDMS blends with 0.5–3.0 wt % of PDMS could be melt-spun into fibers well even when the melt temperature has a drop of 2–3 °C. Especially when PDMS addition is 0.5–1.0 wt %, the blend showed outstanding spinning capability. Furthermore, when comparing the mechanical properties among all obtained fibers, the addition of 0.5–1.0 wt % PDMS was found optimal. As listed in Table 1, certain enhancement of tensile-strength was observed, which may be ascribed to the role of PDMS that could improve mobility of PA66 chains, lowering entanglement of PA66 macromolecular chains. Under the same drawing ratio, the existence of PDMS can make the blend fiber maintain a high elongation at break, which could lead to a better hand feeling of fiber and textile. Regarding dry heat shrinkage, the test results (Figure 3) gave a declining trend when more PDMS was added, indicating a better performance of dimensional stability under high temperature. The reason for this reduced thermal shrinkage may be ascribed to the presence of PDMS which could decrease intermolecular interaction and internal stress in fiber.

Table 1. Tensile strength and elongation of PA66/PDMS blend fibers with different addition of PDMS at varied drawing ratio.

Sample	Drawing Ratio	Tensile Strength (cN/dtex)	Elongation (%)
PA66	3.5	4.3	28.5
PA66/0.5%PDMS	3.5	4.5	35.2
PA66/1%PDMS	3.5	4.4	36.9
PA66	3.7	4.5	26.9
PA66/0.5%PDMS	3.7	4.7	29.5
PA66/1%PDMS	3.7	4.7	30.4
PA66	3.9	5.3	22.1
PA66/0.5%PDMS	3.9	5.5	23.9
PA66/1%PDMS	3.9	5.4	24.4
PA66	4.1	5.6	18.6
PA66/0.5%PDMS	4.1	5.8	18.9
PA66/1%PDMS	4.1	5.7	19.1

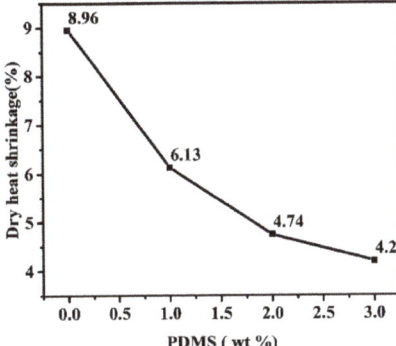

Figure 3. Correlation between dry heat shrinkage and PDMS content.

3.4. Surface Properties of PA66/PDMS Blend Fibers

EDS was applied to detect whether silicon existed on the surface or cross section of PA66/PDMS blend fibers. Figure 4a,c illustrates the surface and cross section morphology of PA66/PDMS (3%) blend fiber with 3 wt % of PDMS, while its corresponding silicon mapping can be found on Figure 4b,d. It is noticed that silicon exists on both surface and cross section of the fiber, also there is a trend showing silicon moving onto the surface. Apart from that, the element content analysis explains that the amount on the surface is higher than inside, ensuring when external silicon is falling off as time passes, the element within can still move and concentrate onto the outer layer. This, in contrast with other surface coating or fishing with silicon agent, could function well in long duration as well.

The existence of silicon on the surface could decrease the value of surface friction coefficient. According to Table 2, the dynamic (μs) and static (μd) friction coefficient of pristine PA66 is 0.3712 and 0.3217, respectively. With the increasing addition of PDMS, both μs and μd experience a downward trend, e.g., comparing pristine PA66 fiber with PA66/1%PDMS fiber, there is a significant drop in both μs and μd values. This finding is consistent with the concentration of PDMS on the fiber surface which tends to become saturated with increasing PDMS addition. A small value of static and dynamic friction coefficient means better behavior in aspects of smoothness and stain-resistance of fibers.

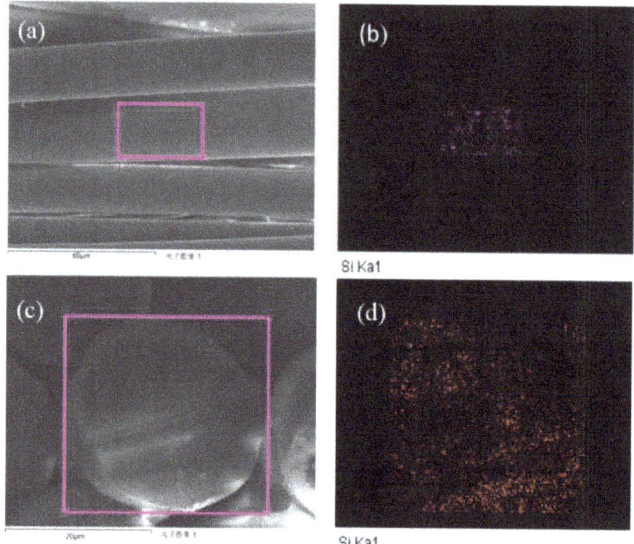

Figure 4. Electron scanning micrograph (a,c) and element mapping of surface and cross section of PA66/3%PDMS blend fiber (b,d).

Table 2. Surface friction coefficient of the obtained fibers.

Sample	fs/10-3cN	μs	fd/10-3cN	μd
PA66	137.5	0.3712	127.0	0.3217
PA66/1%PDMS	110.8	0.2572	105.9	0.2404
PA66/2%PDMS	101.8	0.2276	98.2	0.2155
PA66/3%PDMS	99.8	0.2203	96.8	0.2110

It was reported that when PDMS was used to treat cotton fiber and its fabric for surface coating the directional alignment of PDMS was observed, i.e., the polar silicon–oxygen bond pointed to the surface and the silicon-containing methyl to the air, which together formed a hydrophobic layer [1,10]. We can expect similar results emerge in obtained PA66/PDMS blend fibers. The water contact angle on fiber surface was given in Figure 5, suggesting an increase of contact angle as PDMS increases. A large contact angle could be able to improve waterproof performance, which, together with help from anti-fouling property endowed by small friction coefficient, can add extra value to sportswear et al.

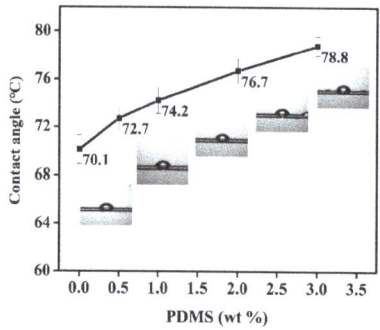

Figure 5. Correlation between PDMS content (%) and contact angle.

3.5. Bending Deformation Properties of PA66/PDMS Blend Fibrous Fabrics

All fabric understands bending deformation during textile processing and practical application [11,12]. The resistance of fabric to this bending is defined as bending rigidity (B), i.e., the ratio of small variety of bending moment per unit width over the change in curvature in response. The smaller the value is, the softer the fabric would be. Combined with elasticity represented by 2HB called bending hysteresis, we can evaluate the general performance of hand feeling and fabric style. 2HB stands for the margin between the curvature per unit width when the sample is forced and the curvature value after recovery when the force is released. A smaller value of 2HB implies larger fabric elasticity and greater bouncing style. Figure 6 is the force–curvature curves of different fabrics, after calculation based on we can find that with incorporation of PDMS in fiber, the blend fibrous fabric exhibits declining tendency of B and 2HB in both wrap and weft (Table 3), suggesting an improvement in softness, elasticity, and bouncing style of fabrics.

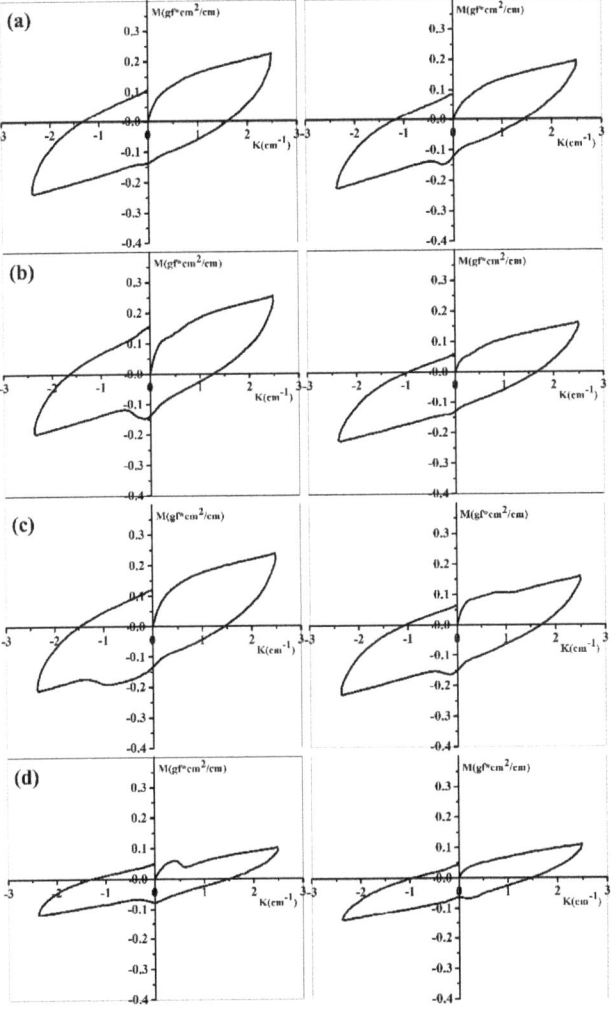

Figure 6. Bending deformation properties of the fabric: (**a**) PA66 in wrap and weft, (**b**) PA66/1%PDMS in wrap and weft, (**c**) PA66/2%PDMS in wrap and weft, (**d**) PA66/3%PDMS in warp and weft.

Table 3. Bending deformation properties of fabrics.

Sample	B (cN·cm²/cm)		2HB (cN·cm/cm)	
	Warp	Weft	Warp	Weft
PA66	0.0549	0.0496	0.2000	0.1656
PA66/1% PDMS	0.0571	0.0447	0.1967	0.1596
PA66/2% PDMS	0.0277	0.0322	0.2191	0.1741
PA66/3% PDMS	0.0283	0.0313	0.0872	0.0931

Basically, the softness, elasticity, and bulkiness of fabric can be interpreted by compression properties in the direction of thickness, whose indexes can be listed as compression linearity (LC), compression work (WC), and elasticity rate of compression work (RC). With reference to Table 4, after addition of PDMS, LC and WC do not express significant change while RC experiences an increasing tendency as PDMS increases. RC represents the recovery ability after the pressure on fabric is released; the larger the value, the better the recovery ability the fabric has. Consistent with the bending deformation test above, the fibrous fabric performs better in elastic recovery properties with presence of PDMS in fiber.

Table 4. Compression properties of the fabrics.

Sample	LC	WC (cN·cm/cm²)	RC (%)
PA66	0.28	0.17	29.4
PA66/1% PDMS	0.25	0.17	31.6
PA66/2% PDMS	0.26	0.15	30.7
PA66/3% PDMS	0.22	0.12	33.3

4. Conclusions

PDMS can be successfully incorporated and well distributed in PA66 matrix fiber through a simple direct melt-mixing and spinning process. The obtained PA66/PDMS blend fiber showed satisfied mechanical properties and reduced thermal shrinkage at elevated temperature, while almost no effect on crystallization and melting behavior of PA66. Furthermore, with addition of PDMS, the decreased surface friction coefficient and increased water contact angle on the PA66/PDMS fiber surface was found, which may provide function of antifouling, waterproof, and stain-resistant ability of the fiber and its fabric. Fabric style measurement by KES serial equipment explained that PA66/PDMS blend fibrous fabric presents improved handing of smoothness, softness, and also a high level of elasticity against compression. Comparing with silicon compound coating or finishing on fiber and fabric surface, the method reported in this article is more eco-friendly and the results proved effective.

Author Contributions: Conceptualization, W.P. and Y.Q.; methodology, T.Z. and W.P.; validation, S.Y., J.J., and G.L.; Formal analysis, S.Y.; Investigation, J.J.; Writing—Original Draft preparation, W.P.; Writing—Review and Editing, Y.Q. and T.Z.; Supervision, G.L.; Project administration, J.J.; Funding acquisition, G.L.

Funding: This work was financially supported by the National Key Research and Development Program of China (2016YFB0303001).

Conflicts of Interest: The authors declare no conflict of interest.

References

1. Owen, M.J.; Evans, J.I. Industrial applications of the surface activity of polydimethylsiloxanes. *Br. Polym. J.* **1975**, *7*, 235–245. [CrossRef]
2. Zięba, M.; Małysa, A.; Wasilewski, T.; Ogorzałek, M. Effects of Chemical Structure of Silicone Polyethers Used as Fabric Softener Additives on Selected Utility Properties of Cotton Fabric. *Autex Res. J.* **2019**, *19*, 1–7. [CrossRef]

3. Kamelian, F.S.; Mohammadi, T.; Naeimpoor, F. Fast, facile and scalable fabrication of novel microporous silicalite-1/PDMS mixed matrix membranes for efficient ethanol separation by pervaporation. *Sep. Purif. Technol.* **2019**, *229*. [CrossRef]
4. Li, S.; Liu, X.; Li, L.; Zhang, H.; Qiu, C. Drag-reductive and anti-corrosive superhydrophobic surface fabricated on aluminum with thin PDMS/SiO$_2$ coating. *Mater. Res. Express* **2019**, *6*. [CrossRef]
5. Hager, R. A Novel Product Concept: Silicone Gum Pekkets as Addutuves for Thermoplasts. *Plast. Addit.* **2004**, *6*, 6.
6. Ji-min, S.U.; Xiao-rui, L.I.; Hai-hua, W.; Gui-qiang, F.E.I. Preparation of Organosilicon Emulsion Stabilized with Hydrophilic Polyether Silicon Oil and Its Defoaming and Foam Suppression Properties. *Trans. China Pulp Pap.* **2009**, *24*, 51–55.
7. Jörn, D.; Kohorst, P.; Besdo, S.; Rücker, M.; Stiesch, M.; Borchers, L. Influence of lubricant on screw preload and stresses in a finite element model for a dental implant. *J. Prosthet. Dent.* **2014**, *112*, 340–348. [CrossRef] [PubMed]
8. Wan, H.; Ye, Y.; Chen, L.; Chen, J.; Zhou, H. Influence of Polyfluo-Wax on the Friction and Wear Behavior of Polyimide/Epoxy Resin–Molybdenum Disulfide Bonded Solid Lubricant Coating. *Tribol. Trans.* **2016**, *59*, 889–895. [CrossRef]
9. Zuber, M.; Zia, K.M.; Tabassum, S.; Jamil, T.; Barkaat-ul-Hasin, S.; Khosa, M.K. Preparation of rich handles soft cellulosic fabric using amino silicone based softener, part II: Colorfastness properties. *Int. J. Biol. Macromol.* **2011**, *49*, 1–6. [CrossRef] [PubMed]
10. Zia, K.M.; Tabassum, S.; Barkaat-ul-Hasin, S.; Zuber, M.; Jamil, T.; Jamal, M.A. Preparation of rich handles soft cellulosic fabric using amino silicone based softener. Part-I: Surface smoothness and softness properties. *Int. J. Biol. Macromol.* **2011**, *48*, 482–487. [CrossRef] [PubMed]
11. Murugan, R.; Ramesh, R.; Padmanabhan, K. Influence of Stacking Sequence on Mechanical Properties and Vibration Characteristics of Glass/Carbon Hybrid Plates with Different Fabric Areal Densities. *Struct. Integr. Assess.* **2019**, 87–97. [CrossRef]
12. Katiyar, P.; Mishra, S.; Srivastava, A.; Prasad, N.E. Preparation of TiO$_2$-SiO$_2$ Hybrid Nanosols Coated Flame-Retardant Polyester Fabric Possessing Dual Contradictory Characteristics of Superhydrophobicity and Self Cleaning Ability. *J. Nanosci. Nanotechnol.* **2020**, *20*, 1780–1789. [CrossRef] [PubMed]

© 2019 by the authors. Licensee MDPI, Basel, Switzerland. This article is an open access article distributed under the terms and conditions of the Creative Commons Attribution (CC BY) license (http://creativecommons.org/licenses/by/4.0/).

Article

Synergistic Effects of Two-Dimensional MXene and Ammonium Polyphosphate on Enhancing the Fire Safety of Polyvinyl Alcohol Composite Aerogels

Xinxin Sheng [1,*], Sihao Li [1], Yanfeng Zhao [1,2], Dongsheng Zhai [1], Li Zhang [1] and Xiang Lu [3,*]

1. Guangdong Provincial Key Laboratory of Functional Soft Condensed Matter, School of Materials and Energy, Guangdong University of Technology, Guangzhou 510006, China; 2111902041@mail2.gdut.edu.cn (S.L.); 2017021930@m.scnu.edu.cn (Y.Z.); dongshengzhai567@126.com (D.Z.); lizhang@gdut.edu.cn (L.Z.)
2. School of Chemistry and Environment, South China Normal University, Guangzhou 510006, China
3. Key Laboratory of Polymer Processing Engineering of the Ministry of Education, National Engineering Research Center of Novel Equipment for Polymer Processing, South China University of Technology, Guangzhou 510641, China
* Correspondence: xinxin.sheng@gdut.edu.cn (X.S.); luxiang_1028@163.com (X.L.)

Received: 6 November 2019; Accepted: 27 November 2019; Published: 29 November 2019

Abstract: Fire and smoke suppressions of polyvinyl alcohol (PVA) aerogels are urgently required due to the serious fire hazard they present. MXene, a 2D transition-metal carbide with many excellent properties, is considered a promising synergist for providing excellent flame retardant performance. PVA/ammonium polyphosphate (APP)/transition metal carbide (MXene) composite aerogels were prepared via the freeze-drying method to enhance the flame retardancy. Thermogravimetric analysis, limiting oxygen index, vertical burning, and cone calorimeter tests were executed to investigate the thermal stability and flame retardancy of PVA/APP/MXene (PAM) composite aerogels. The results demonstrated that MXene boosted the flame retardancy of PVA-APP, and that PAM-2 (with 2.0 wt% MXene loading) passed the V-0 rating, and reached a maximum LOI value of 42%; Moreover, MXene endowed the PVA-APP system with excellent fire and smoke suppression performance, as the the peak heat release rate and peak smoke production rate were significantly reduced by 55% and 74% at 1.0 wt% MXene loading. The flame retardant mechanism was systematically studied, MXene facilitated the generation of compact intumescent residues via ita catalyst effects, thus further restraining the release of heat and smoke. This work provides a simple route to improve the flame retardancy of PVA aerogels via the synergistic effect of MXene and APP.

Keywords: MXene; synergistic effect; ammonium polyphosphate; polyvinyl alcohol (PVA)

1. Introduction

Aerogels are porous solid materials composed of nanoscale colloidal particles or polymer materials [1]. Their unique open nanoscale porous structure and continuous three-dimensional network structure give them significantly low density, high specific surface area and high porosity, wherein the solid phase of the aerogel accounts for only about 0.2–20% of the total volume. Due to the strong adsorption ability, low thermal conductivity, low acoustic impedance and low refractive index, etc., most aerogels have broad application prospects in aviation and aerospace, chemical industry, metallurgical industry, energy-saving buildings, and so on [2,3].

Polyvinyl alcohol (PVA) is a water-soluble and very biocompatible polymer material [4,5], which is often used to prepare organic aerogels by freeze-drying. Compared to inorganic aerogels, PVA organic aerogels usually exhibit better fracture toughness. However, PVA is very easily ignited and has a quite fast fire spread speed owing to its organic structure, and its LOI is only around 20.0%. The flammability of PVA aerogels restricts their effective application in some fields that require high flame retardancy, thus it is very essential to improve the flame retardancy of PVA aerogels.

In general, the incorporation of flame retardants (FR) into polymer materials is the most effective method to lower the flammability of polymer-based materials. Among all FRs, intumescent flame retardants (IFRs), such as ammonium polyphosphate (APP), are considered to be some of the most promising, eco-friendly, and halogen-free flame retardants thanks to the non-toxicity and anti-dripping properties during the combustion process. However, the traditional IFRs have bad flame retardant efficiency and thermal stability. Compared to halogenated flame retardants, higher mass fractions of halogen-free flame retardants are required to achieve the same flame retardant rating for the polymer-based materials, and the use of a great quantity of halogen-free flame retardant not only causes a decrease in the physical properties of the resulting material, but also causes an increase in material cost. In order to solve these problems, synergists are incorporated to enhance the flame retardancy of IFR and reduce the amount of the IFR required in the polymer matrix.

Li et al. [6] prepared a series of zeolites (ZEO) and APP synergistic flame retardant polyurethane foams (PUFs), and investigated the effect of ZEO on the mechanical performance, thermal performance, flame retardant properties of the PUF/APP composites; the results demonstrated that, with the introduction of ZEO, the flame retardancy of PUF/APP composites was obviously improved, and a synergistic effect between ZEO and APP was observed. Liu et al. [7] found that there is obviously synergistic effect between lanthanum oxide (La_2O_3) and IFR for polylactide (PLA)/IFR composites, as the flame retardant performance indices, such as limited oxygen index (LOI) and vertical burning (UL-94), of the PLA/IFR composite were dramatically enhanced with the addition of La_2O_3. Feng et al. [8] synthesized a series of maleated cyclodextrins (MCs) and their metal salts (metal MCs), and used them as the synergists for PVA/IFR composites. The results show that the metal MCs could facilitate the generation of an organized and dense char layer in the condensed phase during the degradation of PVA, and the thermal stability and vertical burning level of the PVA/IFR composites was greatly improved. All the above studies show that the addition of a small amount of synergist to a polymer/IFR composite is an effective way to significantly improve the flame retardant properties of the polymer/IFR composite.

MXene is a new type of two-dimensional (2D) layered transition metal carbide/carbonitride, where M represents early transition metal and X stands for carbon and/or nitrogen. $Ti_3C_2T_x$ (T represents the surface termination, including −O, −OH, or −F) is a kind of widely studied MXene, due to its good electromagnetic, electronic conductivity, chemical stability, antibacterial properties [9–11]. Sheng et al. [12] proved that the incorporation of exfoliated MXene nanosheets resulted in the significant enhancement on the mechanical and thermal performances of thermoplastic polyurethane (TPU). Wang et al. [9] fabricated an annealed $Ti_3C_2T_x$ MXene/epoxy nanocomposite, and found that it exhibited an excellent EMI shielding effectiveness (SE) and electrical conductivity. Yu et al. [13] revealed that the addition of cetyltrimethylammonium bromide (CTAB)- and tetrabutylphosphine chloride (TBPC)-functionalized MXene nanomaterials could remarkably enhance the fire safety and smoke suppression properties of TPU. However, the strategy for the fabrication of MXene-based flame retardant polymers has the drawback of complex manipulation, and it is thus difficult to industrialize the production.

To date, there have been no reports on the synergistic flame retardant effect of MXene on the polymer/halogen-free flame-retardant system. MXene is regarded as a new class of the 2D nanocarbon-based material, which has important theoretical and practical significance for its synergistic effect with halogen-free flame retardants on the flame retardant effect of polymer matrix. Hence, in this work, MXene was prepared and the effect of synergy between MXene and APP on flame retardancy of PVA aerogel was investigated. LOI, UL-94, cone calorimeter tests, and the thermal stability of all samples were investigated. The results showed that the inclusion of MXene can further enhance the flame retardancy of PVA-APP composite aerogels, providing excellent fire and smoke suppression properties.

2. Materials and Methods

2.1. Materials

Polyvinyl alcohol (PVA) with a degree of polymerization of 1700 and a saponification degree of 99 mol %, lithium fluoride (LiF), and lithium chloride (LiCl), were purchased from Aladdin Chemicals Co., Ltd. (Shanghai, China). Hydrochloric acid (HCl, 37 wt%) was obtained from Guangzhou Chemical Reagent Factory (Guangzhou, China). Ammonium polyphosphate (APP) was purchased from Clariant Chemicals Co., Ltd. (Guangzhou, China) MAX phase (Ti_3AlC_2 99.7%, 300 mesh) powder was purchased from Nanjing Mission Advanced Materials Co. (Nanjing, China), deionized water was made in the laboratory. All the reagents were used as received.

2.2. Synthesis of the $Ti_3C_2T_x$ (MXene) Nanosheets

The $Ti_3C_2T_x$ (MXene) nanosheets were synthesized as followed: firstly, 2.0 g of MAX powder was slowly loaded into 40 mL of 3 M LiF and 9 M HCl. Then, the mixture of layered MXene and unetched MAX phase was prepared by continuous stirring at 38 °C for 48 h. The above solutions were washed separately three times with 1 M HCl and LiCl via centrifugation. Subsequently, the solution was centrifuged and washed several times with deionized water until the pH for the supernatant was washed approach to neutral. The precipitates collected by centrifugal washing were distributed in the flask with deionized water and protected with nitrogen for 20 min. The sediment was redispersed with an additional 100 mL of deionized water followed by mixed ultrasonically at ice-bath for 1 h under nitrogen atmosphere. Finally, the solution was then centrifuged for 60 min at 3500 rpm, to give the supernatant slurry containing the exfoliated MXene nanosheets.

2.3. Fabrication of PVA-APP-MXene Composite Aerogels

Before processing, PVA and APP were dried in an oven at 80 °C for 24 h. The preparation process for PVA-APP-MXene aerogel was exhibited in Scheme 1. The formulation of PVA composite aerogels was shown in Table 1. A sample containing 85 wt% PVA, 14.5 wt% APP and 0.5 wt% MXene is denoted as PAM-0.5 (where P stands for PVA, A stands for APP and M stands for MXene). In a typical preparation process of PAM-0.5, 85.0 g of PVA powder and 14.5 g of APP were mixed in a certain amount of deionized water and stirred at 85 °C. 0.5 g of net MXene (in the form of colloidal dispersion) were then added to the mixture; it should be noted that the solid content of the mixture was kept at 10 wt%, and mechanically stirred for 4 h at a speed of 200 rpm under a nitrogen atmosphere. Finally, the mixture was poured into the mold to be frozen into ice, following by freeze-dried in a freeze-dryer under a high vacuum (<8 Pa). Before test, all samples were maintained at 80 °C for 24 h under vacuum for desiccation.

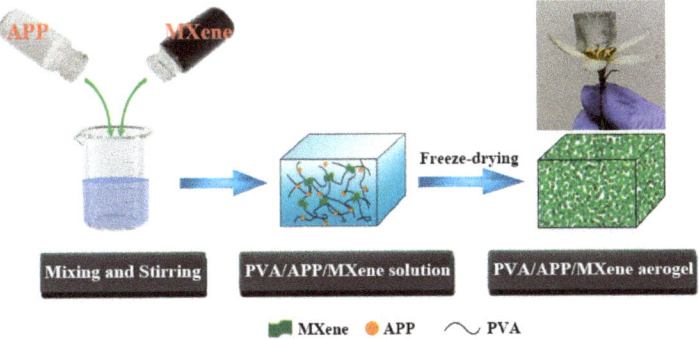

Scheme 1. Preparation process for PVA/APP/MXene composite aerogel.

Table 1. Formulation and apparent density of pure PVA and its composite aerogels.

Samples	PVA (wt%)	APP (wt%)	MXene (wt%)	Apparent Density [1] (mg/cm^3)
PVA	100	-	-	124.4 ± 0.9
PAM-0	85	15	-	127.7 ± 0.8
PAM-0.5	85	14.5	0.5	126.0 ± 1.1
PAM-1	85	14.0	1.0	121.3 ± 0.7
PAM-2	85	13.0	2.0	120.8 ± 0.8

[1] Apparent density was calculated from the quality and volume of the aerogel.

2.4. Characterization

The microstructure of all samples was determined with a SU8010 scanning electronic microscope (HITACHI, Tokyo, Japan) at 15 kV. The thickness and lateral sizes of MXene were measured with a Multimode Nano 4 atomic force microscope (Bruker, Karlsruh, Germany) in tapping-mode. Intercalated and exfoliated structures of MXene were performed with a D/Max-2500 X-ray diffractometer (Rigaku, Tokyo, Japan) using Cu-Kα radiation (λ = 0.15406 nm). The thermal degradation behavior of all aerogels was executed with a STA449F3 thermogravimeter (NETZSCH, Selb, Germany) from 40 °C to 600 °C with a 10 °C min^{-1} heating rate under 100 mL min^{-1} a nitrogen flow. LOI values were performed with a COI oxygen index meter (Mordis, Kunshan, China) with a dimension of 100 mm × 10 mm × 3 mm according to ASTM D2863-97. The vertical burning test (UL-94) was performed with a CZF-6 instrument (Jiangning, Nanjing, China) using samples with dimensions of 130 mm × 13 mm × 3.2 mm according to ATSM D 3801. Fire behavior was performed with a cone calorimeter device (Vouch, Suzhou, China) with sample dimensions of 100 mm × 100 mm × 3 mm in accordance with ISO 5660 standard processes under a heat flux of 50 kW/m^2. Raman spectra were recorded with a Confocal Raman Microprobe (Renishaw, London, UK) with a wide scope from 600 to 2000 cm^{-1}. The elemental composition of residue was executed with an ESCALAB 250XI spectrometer (Thermo Scientific, Waltham, MA, USA) by using Al-Kα excitation radiation.

3. Results and Discussion

3.1. Characterization of MXene

Figure 1a shows a SEM image of the MAX phase before etching, which presents tightly stacked flakes and a rough surface. Obviously, after in situ hydrofluoric acid etching, a loosely packed clay-like structure could be observed, which conducive to exfoliating of the MXene sheets owing to the weakened interlayer interactions. It can be known from Figure 1c that the main XRD peak (002) of MAX phase at 2θ = 9.61°, which corresponds to a d-spacing of 9.20 Å. The (002) peak of MXene became broadened and shifted to 5.82° (d = 15.17 Å). The result indicates that the layer spacing between MXene layers can be remarkably increased, due to the intercalation of water molecules and lithium ions (Li$^+$) [14]. Moreover, the peak at 2θ = 39° of MXene disappears, further confirming the successful removal Al layers from the MXA phase. The AFM image in Figure 1d exhibits the few-layer structure of the MXene nanosheet with a thickness of 1.5–1.8 nm and lateral sizes up to 600 nm (as shown in Figure 1e,f), which means that it is an ultra-thin 2D material. Hence, the distinctive structure endows the MXene with the capability to uniform dispersed in polymer system.

The obtained cylindrical PVA composite aerogel can stand upright on the flower without damaging its petals (Scheme 1), indicating that the aerogel is quite light in weight. The density of pure PVA and its composite aerogels has an apparent density between 120.8 and 127.7 mg/cm^3 (Table 1). Besides, the microstructures of the PVA, PAM-0 and PAM-1 are presented in Figure 2. As shown in Figure 2a, the PVA exhibits a three-dimensional (3D) architecture with an asymmetrical pore distribution. The pore structure is destroyed upon incorporating of APP (Figure 2b), which may be attributed to the poor compatibility of APP fillers with PVA matrix. As can be seen Figure 2c, it can be observed that the addition of MXene results in a more complete 3D pore network structure and a larger specific surface area presumably because of the sustainable effect of the MXene and the strong hydrogen bonding interactions between MXene and PVA [15].

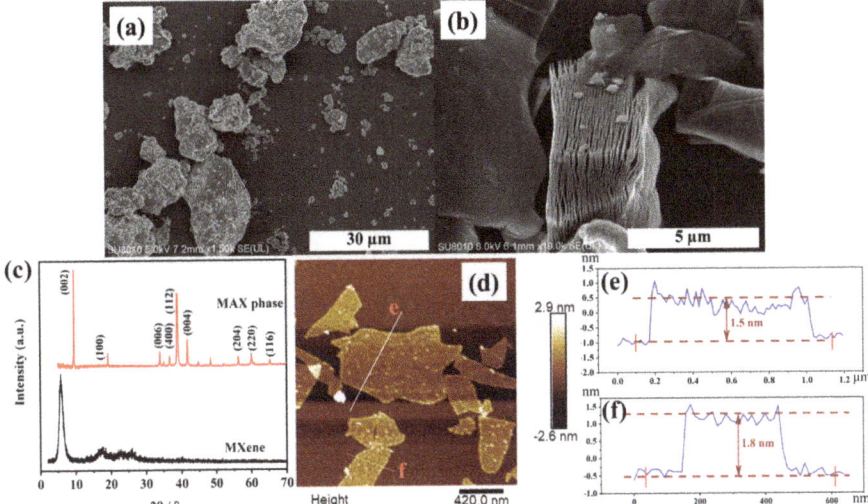

Figure 1. (a) SEM image of MAX phase; (b) SEM image of non-stripped MXene in low resolution; (c) XRD image of MAX phase and MXene; (d–f) AFM image of stripped MXene and the corresponding height profiles.

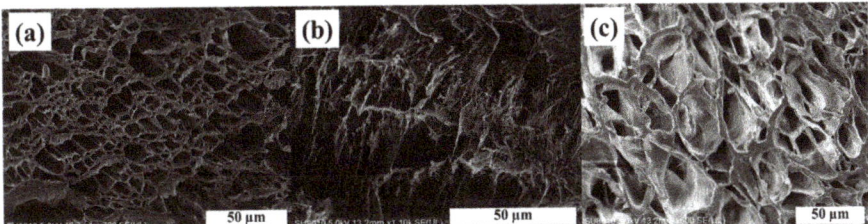

Figure 2. Morphologies of the PVA (a), PAM-0 (b), PAM-1 (c).

3.2. Thermal Degradation Properties

The thermal stability of PVA and its composite aerogels were investigated by TGA under nitrogen. Figure 3 displays thermogravimetric tests results, and the relevant data are summarized in Table 2. Three-stage degradation processes for pure PVA and its composite aerogels are observed in Figure 4b. The first stage from below 100 °C is assigned to the degradation of release of water from the aerogel [16]. The main mass loss stage, between 180 and 400 °C, is ascribed to the degradation of the main chains of PVA. The final mass loss stage is range from 400 to 480 °C, which is ascribed to the further pyrogenic decomposition of the unstable residues created in the main stage. The PVA including APP or a combination of MXene and APP exhibit lower initial degradation temperatures ($T_{5\%}$) and maximum degradation temperatures (T_{max-1}) than the pure aerogel. This may be ascribed to two reasons: on the one hand, the existence of APP promotes the thermal decomposition of the aerogels [17]; on the other hand, the decomposition of MXene surface termination groups [13]. The sample containing MXene show a higher residue at 600 °C than the PAM-0 sample, which implying the introduction of MXene promotes charring and partial polymer cannot be completely burned, leading to enhanced the char residues. Furthermore, in Figure 3b and Table 2, the maximum mass loss rates of PAM-1 composite aerogel is much lower than that of PAM-0. This indicates that MXene inhibits the decomposition of the polymer and promotes the rapid carbon formation of the PVA-APP system, suppressing the diffusion of heat and fuel, thereby enhancing the flame-retardancy of the aerogels.

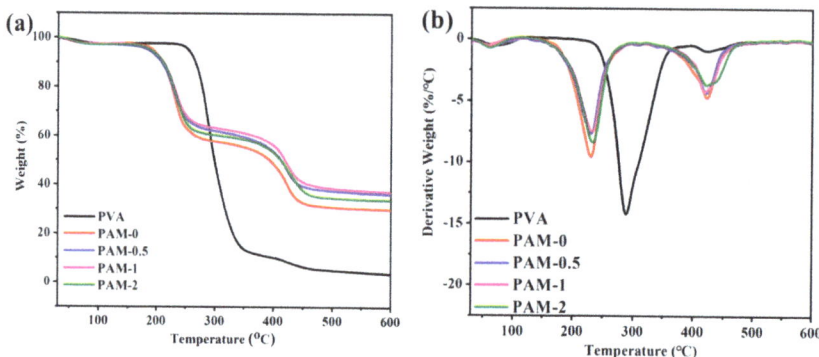

Figure 3. (a) TGA and (b) DTG thermograms of pure PVA and its composite aerogels in nitrogen atmosphere.

Table 2. TGA results for all samples in nitrogen.

Sample	$T_{5\%}$ (°C)	T_{max-1} (°C)	T_{max-2} (°C)	Residue at 600 °C (wt%)
PVA	254	288	424	3.7
PAM-0	192	230	422	29.8
PAM-0.5	183	230	420	36.0
PAM-1	189	232	417	37.0
PAM-2	188	233	421	33.6

3.3. Fire Behavior

The LOI and UL-94 tests of the PVA composite aerogels were conducted, and the corresponding results are displayed in Table 1. PVA aerogel is quickly combustible for its LOI value only 20.5%, and failed the UL-94 test. The inclusion of 15 wt% APP to PVA aerogel result in a strongly improve LOI value (+17%). In the flame retardant PVA systems, partial substitution of APP by MXene leads to further improves in LOI. The LOI values are 39, 41, and 42%, corresponding to the samples with 0.5, 1, and 2 wt% MXene content, respectively. Figure 4 shows digital photographs of the residues of pure PVA and the composite aerogels after UL-94 test.

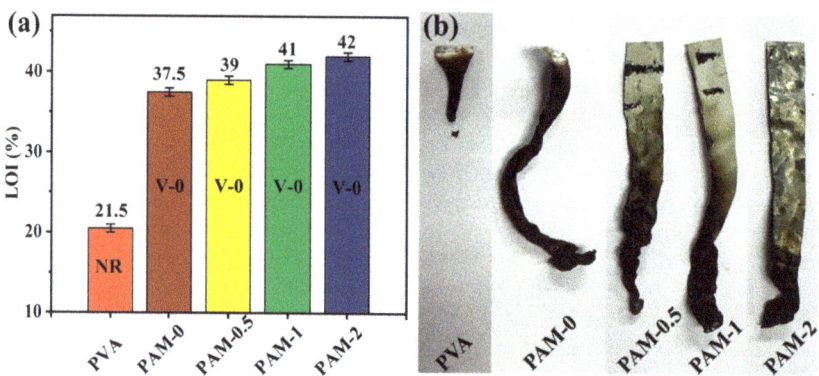

Figure 4. (a) LOI, UL-94 test results and (b) the photos of the char residues after UL-94 test for pure PVA and its composite aerogels.

As observed, the pure PVA is a flammable polymer with serious dripping. All samples with added APP/MXene reached a V-0 rating in the UL-94 test. It can be observed that intumescent residues are formed at 15 wt% APP loading. Furthermore, with the incorporation of MXene, the degree of combustion of the sample during vertical combustion is reduced and the residue is more compact compared with PAM-0. Therefore, MXene can effectively enhance charring of the PVA-APP system. This phenomenon is also in accordance with the results of the previous TGA test.

The fire behaviors of PVA and its composites are estimated by the cone calorimeter. Important fire-hazard parameters containing time to ignition (TTI), time to peak heat release rate (t_{pHRR}), the peak heat release rate (pHRR), total heat release (THR), peak smoke production rate (PSPR), total smoke production (TSP), the maximum average rate of heat emission (MARHE), the effective heat of combustion (EHC), and mass loss are tested. Figure 5 shows HRR, THR, PSPR, TSP, and the mass loss curves of all samples at a heat flux of 50 kW/m^2 and Table 3 summarizes the related data.

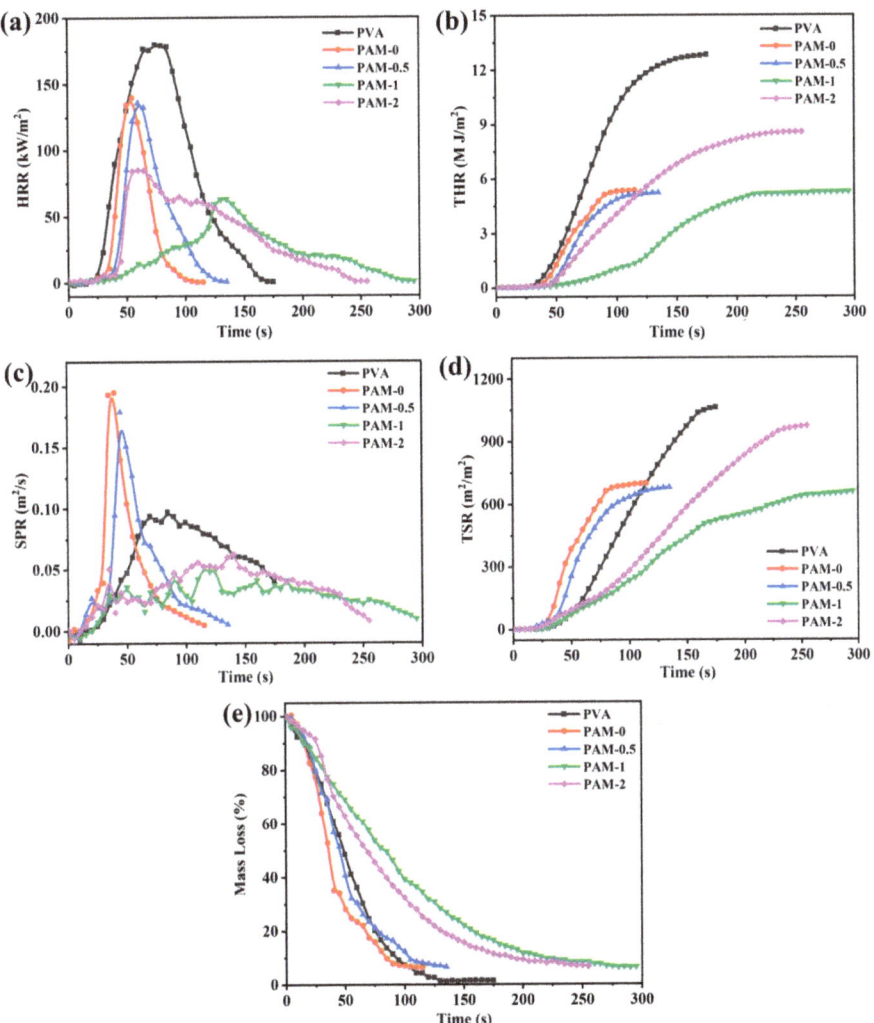

Figure 5. HRR (**a**), THR (**b**), SPR (**c**), TSP (**d**), and mass loss (**e**) as a function of the burning time for PVA and its composite aerogels.

Table 3. Cone calorimeter results of the PVA and its composite aerogels.

Sample	TTI (s)	t_{pHRR} (s)	pHRR (kW/m^2)	THR (MJ/m^2)	MARHE (kW/m^2)	Mean-EHC (MJ/kg)	[a] FPI ((m^2·s)/kW)	PSPR (kW/(m^2·s))	TSR (m^2)	Residue (wt%)
PVA	13	75	180	12.8	99.4	8.96	0.07	0.10	1035	1.4
PAM-0	22	55	140	5.3	50.7	9.55	0.16	0.19	682	6.22
PAM-0.5	26	60	135	5.2	49.2	9.58	0.19	0.18	655	6.5
PAM-1	14	135	63	5.3	32.4	3.87	0.22	0.05	646	6.2
PAM-2	50	60	85	8.5	45.1	9.35	0.59	0.06	945	6.5

[a] FPI = TTI/pHRR.

Figure 5a,b shows the HRR and THR curves of all samples, respectively. Pure PVA aerogel is rapidly burned after ignition, with highest values of pHRR and THR of 180 kW/m^2 and 12.8 MJ/m^2, respectively. Compared with pure PVA aerogel, the pHRR values of the APP-containing composite aerogel decreases to 140 kW/m^2. Moreover, the pHRR values is further declined when the APP is replaced by MXene. The pHRR values of PAM-0.5, PAM-1, PAM-2 are reduced by 3.6%, 55%, 39.3%, respectively, in comparison with PAM-0 sample. Notably, the HRR curves exhibit a flat progress when the loading amount of MXene exceeds 0.5 wt%, illustrating that the aerogels with APP and MXene burnt more slowly. This suggests that MXene could reduce the HRR of PVA-APP system. However, further increasing the MXene loading was unfavorable to the reduction in the PHRR. The results demonstrate that only at the optimum loading MXene can achieve the flame retardant synergistic enhancement effect. The lower or higher loading of MXene has an antagonism effect on the APP and the THR value of PAM-0.5 is further reduced to 5.2 MJ/m^2, which suggests that part of PVA matrix is incompletely burned. Furthermore, the incorporation of APP or MXene can delay the t_{pHRR}. Especially, the t_{pHRR} of the sample with introduction of 14.0 wt% APP and 1.0 wt% MXene reached 135 s, which is greatly prolonged by 60 s compared with the t_{pHRR} of pure PVA, which can increase the escape time and the probability of escape in case of fire.

SPR and TSR curves of all samples are displayed in Figure 5c,d, respectively. The TSR value of PAM-0 is drastically reduced from 1035 m^2/m^2 for pure PVA to 682 m^2/m^2, which corresponds to a 34% reduction. However, the PSPR value of PAM-0 is higher than that of pure PVA, which may be due to the decomposition of APP that can release massive amounts of NH$_3$ and other gases during the burning procedure [18]. Compared with pure PVA, the PSPR and TSP of PAM-1 decreased by 50% and 38%, respectively. The results confirm that MXene can effective enhance smoke production suppression during burning of PVA-APP systems. Hence, the likelihood of suffocation is reduced during the fire escape.

Figure 5e shows the curves of mass loss versus burning time, and the residue yield is shown in Table 3. The mass of pure PVA decreased rapidly, suggesting that the sample is burned completely. while the PAM-0 sample shows a faster decline in mass loss, indicating that APP can promote char formation, but 15 wt% APP is not enough to hinder the combustion of PVA. At the same time, the burning rate is reduced with the introduction of different contents of MXene, of which PAM-1 has the slowest burning rate, indicating that 1.0 wt% MXene can best delay the burning of the system. This is accordance with the heat and smoke release results.

The maximum average rate of heat emission (MARHE) and the fire performance index (FPI) are used to evaluate the fire hazards of aerogels [19,20]. In general, higher FPI and lower MARHE values mean superior fire safety of polymers. The FPI values of PVA, PAM-0, PAM-0.5, PAM-1, PAM-2 were 0.07, 0.16, 0.19, 0.22, 0.59, respectively. It can be observed that the loading of MXene significantly enhanced the FPI value of the system. The MARHE value of PAM-1 is 32.4 kW/m^2, which is the lowest among all samples. Both FPI and MARHE results suggest that the combination of MXene and APP can be highly efficient to improve the fire safety of PVA.

To highlight the significant effect in fire safety and smoke suppression performance of polymers by MXene, a comparative performance evaluation of PVA composite aerogel results of reported MXene-based polymers and PVA containing other flame retardants was performed, as shown in Table 4.

Obviously, although the addition of a single MXene or modified MXene can reduce the PHRR of the system, the THR is reduced slightly [13,21]. MXene/chitosan nanocoating is conductive to suppressing the release of heat and smoke in TPU via the layer-by-layer (LbL) approach, but there is drawback in the complex manipulation [22]. Moreover, traditional nanomaterials can effectively improve the flame retardant performances of the system, however, there they have defects such as large addition or the need for surface modification of the nanomaterials [23–27]. In this work, with our simple strategy without any chemical modification, the introduction of MXene and APP to PVA leads to a significant decrease in PHRR (55%), THR (33.6%) and PSPR (74%), which are better than majority of results reported in the literature, suggesting the superior flame retardant and smoke suppression performance of the proposed combination.

Table 4. PHRR, THR, PSPR, LOI and UL-94 results of MXene based polymer and PVA composites contained other flame retardant in previous work and this work.

Matrix	Flame Retardant	Content (wt%)	PHRR Reduction (%)	THR Reduction (%)	PSPR Reduction (%)	LOI (%)	UL-94	Ref.
TPU	TBPC-Ti_3C_2	2	52.2	1.5	57.4	-	-	[13]
PS	OTAB-Ti_3C_2	2	26.4	-0.09	-	-	-	[21]
PVA	MXene	1	2.6	25.5	-	-	-	[28]
PUF	MXene/Chitosan	-	57.2	65.5	60.3	-	V-0	[22]
PVA	ZrP-EA	8	42.1	16.7	-	-	-	[29]
PVA	HAC	20	-	-	-	30.7	V-0	[30]
PVA	APP/MMT	14.3/0.7	72.7	56.6	-	30.8	V-0	[25]
PVA	MMT/Gelatin	41.7/16.7	13.0	0.68	-	28	-	[26]
PVA	Clay/APP	41.7/16.7	36.8	-36.7	-	-	-	[27]
PVA	APP/MXene	14/1	55	33.6	74	41.0	V-0	This work

TPU: thermoplastic polyurethane; PS: Polystyrene; PUF: Polyurethane foam; TBPC-Ti_3C_2 and OTAB-Ti_3C_2: Inorganic modification Ti_3C_2; ZrP-EA, HAC: Organically modified α-ZrP; MMT: Montmorillonite.

3.4. Analysis of the Residue

Figure 6 exhibits digital photographs of the residues for all aerogels after the cone calorimeter tests. As shown in Figure 6a little residues are left in the pure PVA sample. The residues of PAM-0.5 and PAM-1 are significantly higher than that of PAM-0, indicating that a low addition of MXene enhances the formation of intumescent residue. Moreover, it can be observed that the PAM-1 sample exhibits a more compact and continuous residue, leading to more efficient protection. To further understand the morphology of the residue microstructure, the morphology of the outer (Figure 7a$_1$–d$_1$) and inner (Figure 7a$_2$–b$_2$) residues for PAM-0 and PAM-1 were examined using SEM under different magnifications. A continuous char structure with obvious holes and cracks is observed in the outer surface of PAM-0 char (Figure 7a). As compared to the PAM-1 sample, where the outer residue exhibits a compact and complete surface with compressed layer (Figure 7). The SEM images of inner residues (Figure 7a$_2$–b$_2$) further confirm that the inner char residues of PAM-1 is more compact than that of PAM-0. The char residues of PAM-1 display a more cohesive and pleated layer on the surface with no holes and cracks. As is known to all, a cohesive and continuous surface of residues is conductive to inhibiting the diffusion of heat and inflammable gases, thus enhancing the fire safety [31].

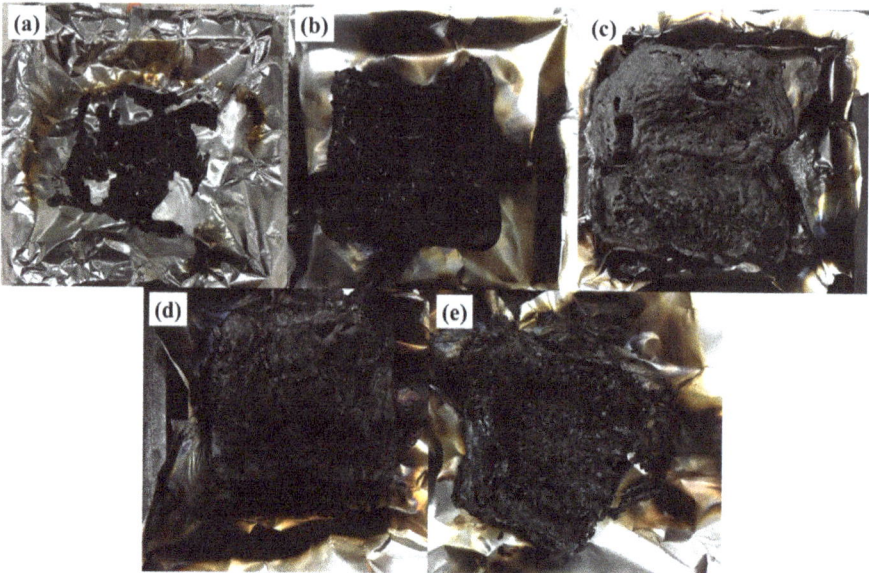

Figure 6. Photographs of residues of (**a**) PVA (**b**) PAM-0, (**c**) PAM-0.5, (**d**) PAM-1, and (**e**) PAM-2 at flame-out after burning.

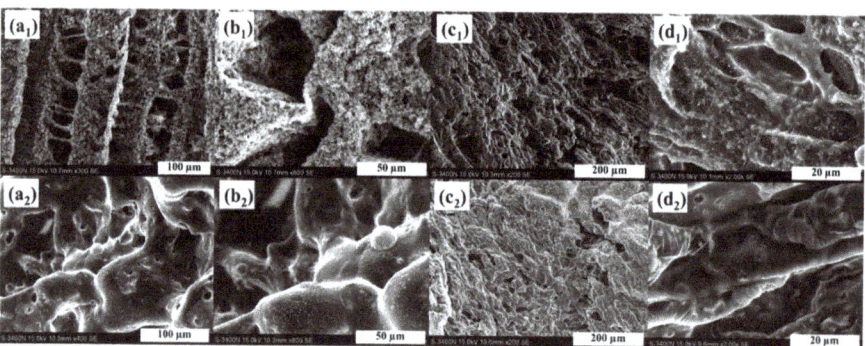

Figure 7. SEM images of the (**1**) outer and (**2**) inner of char residues for PAM-0 (**a**,**b**) and PAM-1 (**c**,**d**).

Raman spectroscopy was performed to investigate the degree of graphitization of the residues after cone calorimetry. Figure 8a–d present the Raman spectra of the residues of the PVA composite aerogels. The degree of graphitization of the residue can be investigated by the intensity ratio of the D (peaks at approximately 1360 cm^{-1}) and G (peaks at approximately 1590 cm^{-1}) bands (I_D/I_G). The carbonaceous material with higher graphitization degree indicates higher thermal oxidation stability, which is conductive to enhancing the flame retardancy of polymers. Generally, a higher graphitization degree corresponds to a lower I_D/I_G value. The incorporation of MXene increases the I_D/I_G value of the PVA-APP system, indicating an increase in the degree of graphitization in the residues.

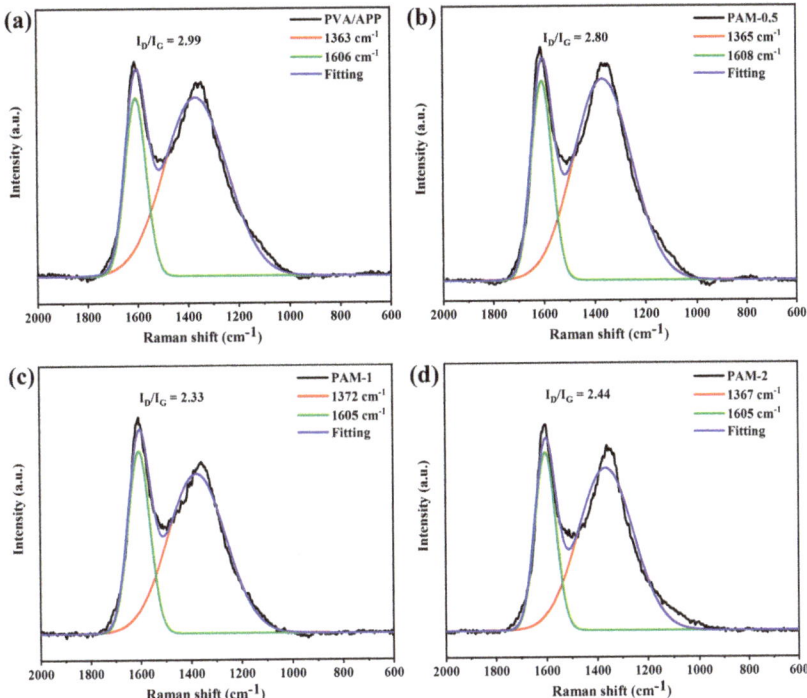

Figure 8. Raman spectra of the char residues of PVA composite aerogels (**a**) PAM-0, (**b**) PAM-0.5, (**c**) PAM-1, and (**d**) PAM-2.

Moreover, the I_D/I_G of PAM-1 is higher than those of PAM-0.5 and PAM-2, suggesting that the residues for PAM-1 as a barrier can have a higher efficiency in preventing heat and fuel diffusion between the condensed phase and the gas phase, thus inhibiting the burning of matrix, and reduce the HRR during burning.

XRD is used to detected the phase composition of the char residues after burning. As shown in Figure 9, it can be observed that a broad diffraction peak at 24.6° is present for the PAM-0, which is ascribed to the crystalline diffraction peak of graphite.

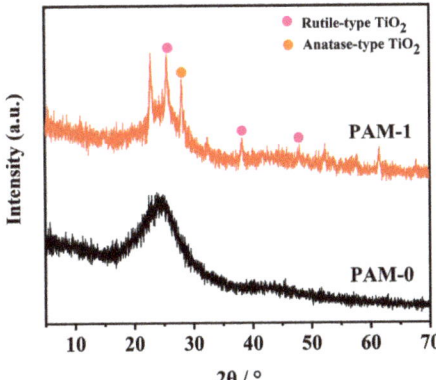

Figure 9. XRD patterns of the char residues of PAM-0 and PAM-1 composite aerogels.

Due to the existence of MXene in the char residue of PAM-1, several new peaks appeared compared to PAM-0. The peaks labelled with purple dot (25.5°, 38.3°, 48.0°) and orange dot (28°) are ascribed to anatase-type and rutile-type TiO_2, respectively [32]. This phenomenon demonstrates that Ti-containing of MXene undergoes thermal oxidation to form stable TiO_2 during burning. Moreover, the TiO_2 may enhance the charring and cross-linking of APP and PVA by catalysis, which further improvez the formation of compact residual char [33].

To further analyze the synergistic mechanism of MXene and APP, XPS was applied to determine the elemental composition and content of PAM-0 and PAM-1. The overall XPS spectra for the residues of the PAM-0 and PAM-1 are shown in Figure 10, and the elemental composition is presented in Table 5. The P/C and N/C ratios of PAM-1 are higher than PAM-0, suggesting an enriched N and P-containing on the residues of PAM-1, which further indicate that the barrier effect of MXene may restrain the release of NH_3 and decrease the volatility of P, and thus more N and P elements are retained in the residues, which is conductive to the formation of more continuous and cohesive residues [22].

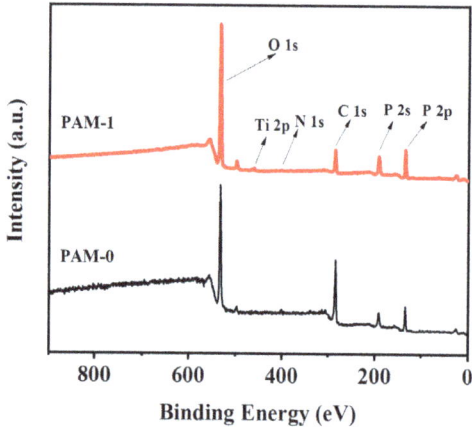

Figure 10. Overall XPS spectra of the char residues after burning in cone calorimeter tests.

Table 5. XPS data of the char residues of PAM-0 and PAM-1.

Sample	C (at%)	O (at%)	N (at%)	P (at%)	Ti (at%)	P/C	N/C
PAM-0	43.90	44.60	2.30	9.20	0	0.21	0.052
PAM-1	19.48	62.44	1.41	15.95	0.72	0.82	0.072

High-resolution XPS spectra for PAM-0 and PAM-1 are presented in Figures 11 and 12. As presented in the C 1s spectrum of PAM-0, the peaks appearing at 284.2, 284.6 and 285 eV correspond to C=C, C–C and C–N bonds. As presented in the O 1s spectrum, the two characteristic peaks at 531.8 and 533 eV are due to C=O/P=O and C–O/P–O–C/P–O–P, implying the formation of crosslinking P-containing structures [31]. The N 1s spectrum is divided into three peaks including C=N, C–N and N–H at binding energies of 398.8, 400.6 and 401.6 eV [34,35], indicating the charring reaction of PVA and APP. In the P 2p spectrum, the signal peak at 134.4 eV is due to P–O–C/PO_3^- in phosphorus-rich crosslinks [36], while for PAM-1, there are three new peaks at 286.3, 288, 289 eV in its C 1s spectrum, which are attributed to C–O, C=O and π–π*, respectively. This suggests that MXene inhibits the decomposition and oxidation of PVA, thereby reducing the release of volatilized gases, such as CO, CO_2, carbonyl compounds, etc [37]. Moreover, the additional element (Ti 2p) signal is split into two peaks: 460 eV (Ti 2p1/2) and 466.5 eV (Ti 2p3/2), which reveals the presence of Ti-containing substances. These suggest that the MXene in PVA-APP system can reduce polymer degradation and the release of volatile gases, thus enhancing the fire safety.

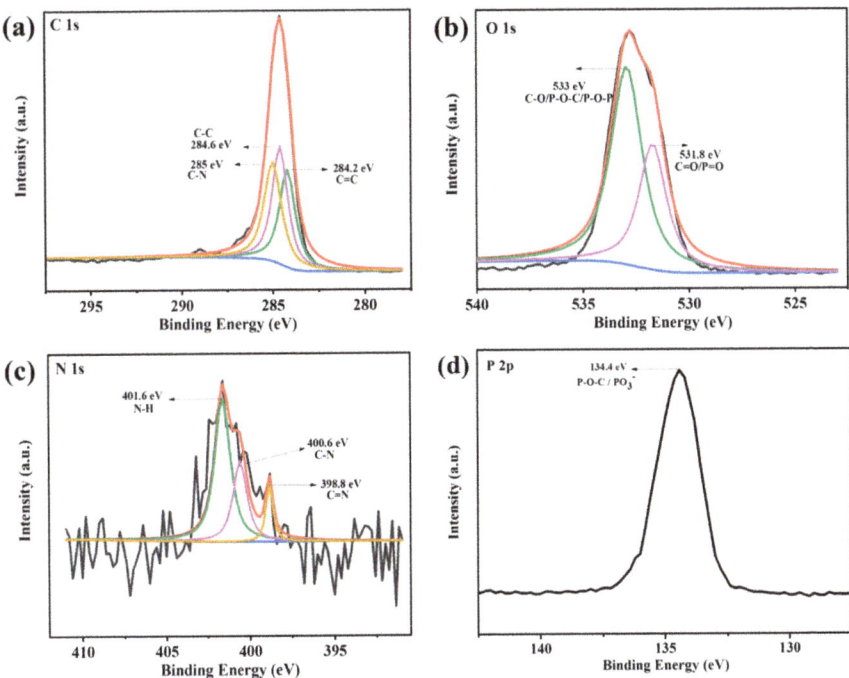

Figure 11. (**a**) C 1s, (**b**) O 1s, (**c**) N 1s, and (**d**) P 2p XPS spectra of the residues for PAM-0.

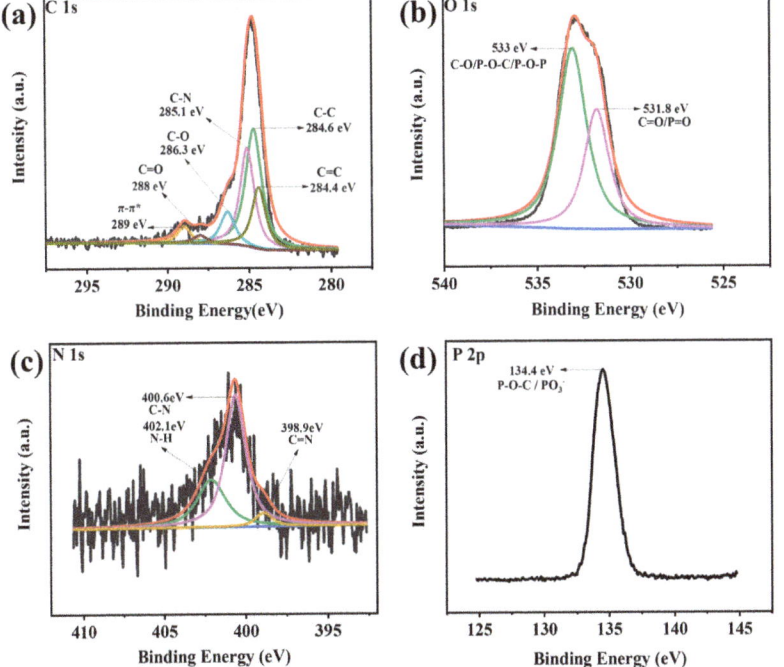

Figure 12. *Cont.*

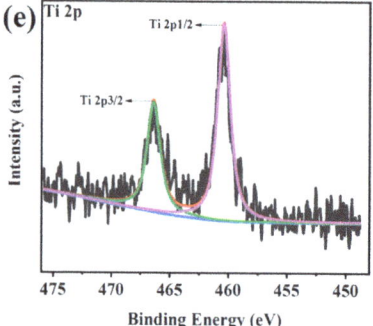

Figure 12. (**a**) C 1s, (**b**) O 1s, (**c**) N 1s, (**d**) P 2p, and (**e**) Ti 2p XPS spectra of the residues for PAM-1.

3.5. Flame Retardant Mechanism

Based on all the above analysis, a possible mechanism of the synergistic interaction between MXene and APP in PVA aerogels is shown in Scheme 2. When the aerogels are ignited, APP begins to decompose from PPA with the release of ammonia (NH_3) and water (H_2O), thus diluting the inflammable gases in the combustion system and formation of intumescent residues, which will decrease the heat release and delay burning [38]. At the same time, a self-crosslinking reaction occurred in part of the PPA. Moreover, PPA may be dehydrated on the PVA chain, which promotes the formation of a stable residue [39]. In this work, MXene would be oxidized on its surfaces and/or edges and generate ultra-small anatase TiO_2 particles due to the strong tendency of the Ti component to react with oxygen during burning, thus consuming oxygen and inhibiting the burning of the matrix. TiO_2 may further boost the cross-linking and charring of APP and/or PVA and promote the formation of compact intumescent residues via catalytic effects [13]. Furthermore, the MXene and APP strengthened the polymer melt viscosity of PVA and enhanced the formation of a compact Ti, P and N hybrid residual char in burning, which as a barrier that inhibits the aerogel from further combustion and restrains the release of smoke and heat [22].

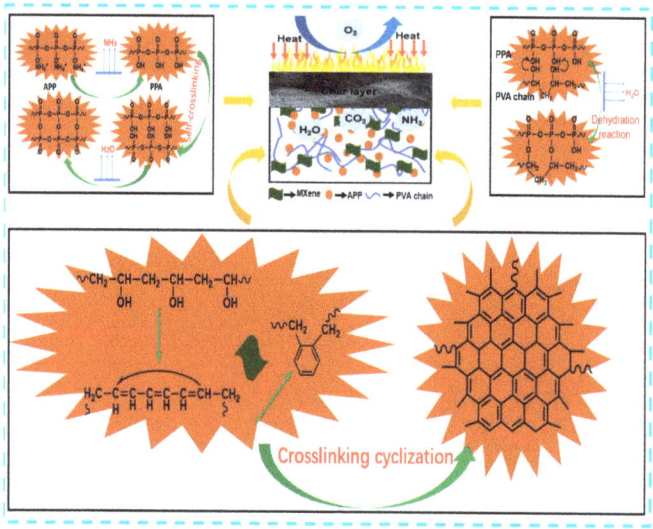

Scheme 2. Schematic illustration for the synergistic flame-retardant mechanism of PVA-APP-MXene composite aerogels.

4. Conclusions

Herein, PVA/APP/MXene composite aerogels were fabricated by a freeze-drying approach. The feasibility of employing MXene as a synergist for APP was investigated in the flame retardancy aspect. With the incorporation of MXene, the fire performance of the PVA aerogels is significantly improved, and adding 1.0 wt% MXene into PVA-APP system leads to a distinct enhancement of the residue yield and a remarkable decrease of the maximum thermal decomposition rate; moreover, MXene can also act as an effective agent for decreasing heat release and suppressing smoke release. In addition, MXene could enhance the degree of graphitization and promote the formation of a continuous char structure. A synergistic flame-retardant mechanism was proposed based on the analyses of the condensed phase residues produced by burning. Furthermore, due to the presence of the MXene, which catalyzes the formation of char, the Ti, P and N hybrid residual char prevents the transform of the heat and fuel, thus improving the flame retardancy of the polymer materials.

Author Contributions: Conceptualization, X.S. and X.L.; Methodology, X.S. and Y.Z.; Validation, S.L., Y.Z. and D.Z.; Formal analysis, S.L.; Investigation, Y.Z.; Writing—original draft preparation, X.S.; Writing—review and editing, X.S. and X.L.; Supervision, L.Z. and X.L.; Funding acquisition, X.S. and X.L.

Funding: This research was funded by the National Natural Science Foundation of China (Grant No. 21908031 and 51903092) and the China Postdoctoral Science Foundation funded project (No. 2019M652884).

Conflicts of Interest: The authors declare no conflict of interest.

References

1. Wang, L.; Schiraldi, D.A.; Sánchez-Soto, M. Foamlike xanthan gum/clay aerogel composites and tailoring properties by blending with agar. *Ind. Eng. Chem. Res.* **2014**, *53*, 7680–7687. [CrossRef]
2. Cuce, E.; Cuce, P.M.; Wood, C.J.; Riffat, S.B. Toward aerogel based thermal superinsulation in buildings: A comprehensive review. *Renew. Sustain. Energy Rev.* **2014**, *34*, 273–299. [CrossRef]
3. Madyan, O.A.; Fan, M. Temperature induced nature and behaviour of clay-PVA colloidal suspension and its aerogel composites. *Colloids Surf. A Physicochem. Eng. Asp.* **2017**, *529*, 495–502. [CrossRef]
4. Tang, X.Z.; Alavi, S. Recent advances in starch, polyvinyl alcohol based polymer blends, nanocomposites and their biodegradability. *Carbohydr. Polym.* **2011**, *85*, 7–16. [CrossRef]
5. Kim, J.O.; Park, J.K.; Kim, J.H.; Jin, S.G.; Yong, C.S.; Li, D.X.; Choi, J.Y.; Woo, J.S.; Yoo, B.K.; Lyoo, W.S. Development of polyvinyl alcohol-sodium alginate gel-matrix-based wound dressing system containing nitrofurazone. *Int. J. Pharm.* **2008**, *359*, 79–86. [CrossRef]
6. Li, J.; Mo, X.; Li, Y.; Zou, H.; Liang, M.; Chen, Y. Effect of zeolites on morphology and properties of water-blown semi-rigid ammonium polyphosphate intumescent flame-retarding polyurethane foam. *J. Polym. Res.* **2017**, *24*, 154. [CrossRef]
7. Feng, C.; Liang, M.; Zhang, Y.; Jiang, J.; Huang, J.; Liu, H. Synergistic effect of lanthanum oxide on the flame retardant properties and mechanism of an intumescent flame retardant PLA composites. *J. Anal. Appl. Pyrolysis* **2016**, *122*, 241–248. [CrossRef]
8. Feng, J.; Zhang, X.; Ma, S.; Xiong, Z.; Zhang, C.; Jiang, Y.; Zhu, J. Syntheses of metallic cyclodextrins and their use as synergists in a poly (Vinyl alcohol)/Intumescent flame retardant system. *Ind. Eng. Chem. Res.* **2013**, *52*, 2784–2792. [CrossRef]
9. Wang, L.; Chen, L.X.; Song, P.; Liang, C.B.; Lu, Y.J.; Qiu, H.; Zhang, Y.L.; Kong, J.; Gu, J.W. Fabrication on the annealed $Ti_3C_2T_x$ MXene/epoxy nanocomposites for electromagnetic interference shielding application. *Compos. Part B Eng.* **2019**, *171*, 111–118. [CrossRef]
10. Rasool, K.; Helal, M.; Ali, A.; Ren, C.E.; Gogotsi, Y.; Mahmoud, K.A. Antibacterial activity of $Ti_3C_2T_x$ MXene. *ACS Nano* **2016**, *10*, 3674–3684. [CrossRef]
11. Kajiyama, S.; Szabova, L.; Sodeyama, K.; Iinuma, H.; Morita, R.; Gotoh, K.; Tateyama, Y.; Okubo, M.; Yamada, A. Sodium-ion intercalation mechanism in MXene nanosheets. *ACS Nano* **2016**, *10*, 3334–3341. [CrossRef] [PubMed]

12. Sheng, X.X.; Zhao, Y.F.; Zhang, L.; Lu, X. Properties of two-dimensional Ti_3C_2 MXene/thermoplastic polyurethane nanocomposites with effective reinforcement via melt blending. *Compos. Sci. Technol.* **2019**, *181*, 107710. [CrossRef]
13. Yu, B.; Tawiah, B.; Wang, L.Q.; Yuen, A.C.Y.; Zhang, Z.C.; Shen, L.L.; Lin, B.; Fei, B.; Yang, W.; Li, A. Interface decoration of exfoliated MXene ultra-thin nanosheets for fire and smoke suppressions of thermoplastic polyurethane elastomer. *J. Hazard. Mater.* **2019**, *374*, 110–119. [CrossRef] [PubMed]
14. Ghidiu, M.; Lukatskaya, M.R.; Zhao, M.Q.; Gogotsi, Y.; Barsoum, M. Conductive two-dimensional titanium carbide 'clay' with high volumetric capacitance. *Nature* **2014**, *516*, 78–81. [CrossRef] [PubMed]
15. Ye, S.B.; Feng, J.C.; Wu, P.Y. Highly elastic graphene oxide-epoxy composite aerogels via simple freeze-drying and subsequent routine curing. *J. Mater. Chem. A* **2013**, *1*, 3495–3502. [CrossRef]
16. Wang, Y.T.; Liao, S.F.; Shang, K.; Chen, M.J.; Huang, J.Q.; Wang, Y.Z.; Schiraldi, D.A. Efficient approach to improving the flame retardancy of poly (vinyl alcohol)/clay aerogels: Incorporating piperazine-modified ammonium polyphosphate. *ACS Appl. Mater. Interfaces* **2015**, *7*, 1780–1786. [CrossRef]
17. Camino, G.; Costa, L.; Trossarelli, L. Study of the mechanism of intumescence in fire retardant polymers: Part II-Mechanism of action in polypropylene-ammonium polyphosphate-pentaerythritol mixtures. *Polym. Degrad. Stab.* **1984**, *7*, 25–31. [CrossRef]
18. Laoutid, F.; Bonnaud, L.; Alexandre, M.; Lopez-Cuesta, J.M.; Dubois, P. New prospects in flame retardant polymer materials: From fundamentals to nanocomposites. *Mater. Sci. Eng. R Rep.* **2009**, *63*, 100–125. [CrossRef]
19. Schartel, B.; Hull, T.R. Development of fire-retarded materials-interpretation of cone calorimeter data. *Fire Mater. Int. J.* **2007**, *31*, 327–354. [CrossRef]
20. Zanetti, M.; Camino, G.; Canavese, D.; Morgan, A.B.; Lamelas, F.J.; Wilkie, C.A. Fire retardant halogen-antimony-clay synergism in polypropylene layered silicate nanocomposites. *Chem. Mater.* **2002**, *14*, 189–193. [CrossRef]
21. Si, J.Y.; Tawiah, B.; Sun, W.L.; Lin, B.; Wang, C.; Yuen, A.C.Y.; Yu, B.; Li, A.; Yang, W.; Lu, H.D. Functionalization of MXene nanosheets for polystyrene towards high thermal stability and flame retardant properties. *Polymers* **2019**, *11*, 976. [CrossRef] [PubMed]
22. Lin, B.; Yuen, A.C.Y.; Li, A.; Zhang, Y.; Chen, T.B.Y.; Yu, B.; Lee, E.W.M.; Peng, S.; Yang, W.; Lu, H.D. MXene/chitosan nanocoating for flexible polyurethane foam towards remarkable fire hazards reductions. *J. Hazard. Mater.* **2019**, *381*, 120952. [CrossRef] [PubMed]
23. Lu, H.D.; Wilkie, C.A. The influence of α-zirconium phosphate on fire performance of EVA and PS composites. *Polym. Adv. Technol.* **2011**, *22*, 1123–1130. [CrossRef]
24. Gu, J.; Liang, C.; Zhao, X.; Gan, B.; Qiu, H.; Guo, Y.; Yang, X.; Zhang, Q.; Wang, D.-Y. Highly thermally conductive flame-retardant epoxy nanocomposites with reduced ignitability and excellent electrical conductivities. *Compos. Sci. Technol.* **2017**, *139*, 83–89. [CrossRef]
25. Lin, J.S.; Liu, Y.; Wang, D.Y.; Qin, Q.; Wang, Y.Z. Poly (vinyl alcohol)/ammonium polyphosphate systems improved simultaneously both fire retardancy and mechanical properties by montmorillonite. *Ind. Eng. Chem. Res.* **2011**, *50*, 9998–10005. [CrossRef]
26. Wang, Y.T.; Zhao, H.B.; Degracia, K.; Han, L.X.; Sun, H.; Sun, M.; Wang, Y.Z.; Schiraldi, D.A. Green approach to improving the strength and flame retardancy of poly (vinyl alcohol)/clay aerogels: Incorporating biobased gelatin. *ACS Appl. Mater. Interfaces* **2017**, *9*, 42258–42265. [CrossRef]
27. Wang, L.; Sánchez-Soto, M.; Maspoch, M.L. Polymer/clay aerogel composites with flame retardant agents: Mechanical, thermal and fire behavior. *Mater. Des.* **2013**, *52*, 609–614. [CrossRef]
28. Pan, Y.; Fu, L.; Zhou, Q.W.; Wen, Z.N.; Lin, C.T.; Yu, J.H.; Wang, W.M.; Zhao, H.T. Flammability, thermal stability and mechanical properties of polyvinyl alcohol nanocomposites reinforced with delaminated $Ti_3C_2T_x$ (MXene). *Polym. Compos.* **2019**. [CrossRef]
29. Lu, H.; Wilkie, C.A.; Ding, M.; Song, L. Thermal properties and flammability performance of poly (vinyl alcohol)/α-zirconium phosphate nanocomposites. *Polym. Degrad. Stab.* **2011**, *96*, 885–891. [CrossRef]
30. Xu, L.F.; Lei, C.H.; Xu, R.J.; Zhang, X.Q.; Zhang, F. Hybridization of α-zirconium phosphate with hexachlorocyclotriphosphazene and its application in the flame retardant poly (vinyl alcohol) composites. *Polym. Degrad. Stab.* **2016**, *133*, 378–388. [CrossRef]

31. Yuan, B.H.; Fan, A.; Yang, M.; Chen, X.F.; Hu, Y.; Bao, C.L.; Jiang, S.H.; Niu, Y.; Zhang, Y.; He, S. The effects of graphene on the flammability and fire behavior of intumescent flame retardant polypropylene composites at different flame scenarios. *Polym. Degrad. Stab.* **2017**, *143*, 42–56. [CrossRef]
32. Li, Z.Y.; Wang, L.B.; Sun, D.D.; Zhang, Y.D.; Liu, B.Z.; Hu, Q.K.; Zhou, A.G. Synthesis and thermal stability of two-dimensional carbide MXene Ti$_3$C$_2$. *Mater. Sci. Eng. B* **2015**, *191*, 33–40. [CrossRef]
33. Feng, X.M.; Xing, W.Y.; Song, L.; Hu, Y.; Liew, K.M. TiO$_2$ loaded on graphene nanosheet as reinforcer and its effect on the thermal behaviors of poly (vinyl chloride) composites. *Chem. Eng. J.* **2015**, *260*, 524–531. [CrossRef]
34. Yuan, B.; Bao, C.; Song, L.; Hong, N.; Liew, K.M.; Hu, Y. Preparation of functionalized graphene oxide/polypropylene nanocomposite with significantly improved thermal stability and studies on the crystallization behavior and mechanical properties. *Chem. Eng. J.* **2014**, *237*, 411–420. [CrossRef]
35. Guo, D.; Wang, Q.; Bai, S. Poly (vinyl alcohol)/melamine phosphate composites prepared through thermal processing: Thermal stability and flame retardancy. *Polym. Adv. Technol.* **2013**, *24*, 339–347. [CrossRef]
36. Deng, C.-L.; Deng, C.; Zhao, J.; Fang, W.-H.; Lin, L.; Wang, Y.-Z. Water resistance, thermal stability, and flame retardation of polypropylene composites containing a novel ammonium polyphosphate microencapsulated by UV-curable epoxy acrylate resin. *Polym. Adv. Technol.* **2014**, *25*, 861–871. [CrossRef]
37. Carja, I.-D.; Serbezeanu, D.; Vlad-Bubulac, T.; Hamciuc, C.; Coroaba, A.; Lisa, G.; López, C.G.; Soriano, M.F.; Pérez, V.F.; Romero Sánchez, M.D. A straightforward, eco-friendly and cost-effective approach towards flame retardant epoxy resins. *J. Mater. Chem. A* **2014**, *2*, 16230–16241. [CrossRef]
38. Shao, Z.B.; Deng, C.; Tan, Y.; Chen, M.J.; Chen, L.; Wang, Y.Z. An efficient mono-component polymeric intumescent flame retardant for polypropylene: Preparation and application. *ACS Appl. Mater. Interfaces* **2014**, *6*, 7363–7370. [CrossRef]
39. Zhao, C.X.; Liu, Y.; Wang, D.Y.; Wang, D.L.; Wang, Y.Z. Synergistic effect of ammonium polyphosphate and layered double hydroxide on flame retardant properties of poly (vinyl alcohol). *Polym. Degrad. Stab.* **2008**, *93*, 1323–1331. [CrossRef]

© 2019 by the authors. Licensee MDPI, Basel, Switzerland. This article is an open access article distributed under the terms and conditions of the Creative Commons Attribution (CC BY) license (http://creativecommons.org/licenses/by/4.0/).

Article

Variation of Ultimate Properties in Extruded iPP-Mesoporous Silica Nanocomposites by Effect of iPP Confinement within the Mesostructures

Rosa Barranco-García [1], José M. Gómez-Elvira [1], Jorge A. Ressia [2,3], Lidia Quinzani [2], Enrique M. Vallés [2], Ernesto Pérez [1] and María L. Cerrada [1,*]

[1] Instituto de Ciencia y Tecnología de Polímeros (ICTP-CSIC), Juan de la Cierva 3, 28006 Madrid, Spain; rbarranco@ictp.csic.es (R.B.-G.); elvira@ictp.csic.es (J.M.G.-E.); ernestop@ictp.csic.es (E.P.)
[2] PLAPIQUI (UNS-CONICET), Camino La Carrindanga km 7, Bahía Blanca 8000, Argentina; jressia@plapiqui.edu.ar (J.A.R.); lquinzani@plapiqui.edu.ar (L.Q.); valles@plapiqui.edu.ar (E.M.V.)
[3] Comisión de Investigaciones Científicas de la Provincia de Buenos Aires (CIC), La Plata 1900, Argentina
* Correspondence: mlcerrada@ictp.csic.es; Tel.: +34-912-587-474

Received: 27 November 2019; Accepted: 20 December 2019; Published: 2 January 2020

Abstract: Nanocomposites based on isotactic polypropylene (iPP) and mesoporous silica particles of either MCM-41 or SBA-15 were prepared by melt extrusion. The effect of the silica incorporated into an iPP matrix was firstly detected in the degradation behavior and in the rheological response of the resultant composites. Both were ascribed, in principle, to variations in the inclusion of iPP chains within these two mesostructures, with well different pore size. DSC experiments did not provide information on the existence of confinement in the iPP-MCM-41 materials, whereas a small endotherm, located at about 100 °C and attributed to the melting of confined crystallites, is clearly observed in the iPP-SBA-15 composites. Real-time variable-temperature Small Angle X-ray Scattering (SAXS) experiments with synchrotron radiation turned out to be crucial to finding the presence of iPP within MCM-41 pores. From these measurements, precise information was also deduced on the influence of the MCM-41 on iPP long spacing since overlapping does not occur between most probable iPP long spacing peak with the characteristic diffractions from the MCM-41 hexagonal nanostructure in comparison with existing superposition in SBA-15-based materials.

Keywords: mesoporous silica; MCM-41 and SBA-15; iPP nanocomposites; thermal degradation; rheological properties; synchrotron SAXS measurements

1. Introduction

Behavior of liquids and solids in very small pores has been a relevant topic, from both a fundamental and practical perspective, for decades [1]. Phenomena, such as the glass transition, phase separation crystallization, and the subsequent melting under confinement, were investigated in order to learn the effect of finite size constraints on bulk properties. This interest also spread to include other materials, such as polymers, with the arrival of nanotechnology in order to attain and understand new physical properties on the molecular scale. The development of polymer-based nanocomposites, thin films and coatings, nanolithography in semiconductor manufacturing, etc. promoted the basic knowledge of the involved molecular phenomena, which are considered critical to the success of the confinement understanding.

Isotactic polypropylene (iPP) is used in a wide variety of applications, such as in the automotive and aerospace industries, because it shows a much desirable versatility and useful physical properties, such as stiffness and strength. Incorporation of specific fillers, leading to the obtainment of micro or nanocomposites, can contribute to the enhancement of some of those excellent properties and to allow spreading out even more its applicability.

Confinement of iPP chains within the ordered spaces present in SBA-15 mesoporous silica was proved recently by Small-angle X-ray scattering (SAXS) using synchrotron radiation, these measurements constituting a reliable and powerful tool [2]. The iPP macrochains filled out those nanometric channels when composites were obtained not only by in situ polymerization [3] but also, unexpectedly, by extrusion [2,4] of the two components, the molten iPP and the SBA-15 particles. Confinement was deduced through the observation of a noticeable discontinuity in the intensity of the first order (100) diffraction of the SBA-15 related to its characteristic hexagonal arrangement. The upward step took place in the temperature interval ranging from 95 to 120 °C, which was fully in agreement with the one noticed for a small endotherm exhibited in the DSC experiments [2–4]. Earlier investigations on several SBA-15 composites [5–7] described that the intensity of the SBA-15 diffraction was dependent on the eventual scattering contrast between walls and inside of the mesopores. These variations were ascribed to those changes in the electron density from iPP chains arranged inside the SBA-15 particles, which were semicrystalline at low temperatures or fully amorphous after the melting of these iPP crystallites existing within mesoporous channels. Moreover, the intensity of that first order reflection was strongly dependent on SBA-15 content [2,4]. These results confirmed the ones described in previous works performed at room temperature on in situ polymerized nanocomposites of poly(N-isopropylacrylamide)/SBA-15 [8] as well as on polyethylenimine (PEI)-based composites wet impregnated with either MCM-41 or SBA-15 [7,9].

Regarding the existence of a small endothermic event in DSC, it should be said that this process was also observed for in situ polymerized materials based on ultra-high molecular weight polyethylene (UHMWPE) and SBA-15 [10,11] as well as in nanocomposites based on high density polyethylene (HDPE) and MCM-41. The temperature range was shifted to lower values [12–14] in the latest ones. These secondary processes were associated with the melting of crystallites with significantly smaller size than the thicker ones that melt at around 130 °C during the main endotherm [1]. These observations seemed to point out that pore size in the silica plays a key role since crystallites cannot grow more than the nanometric spaces where chains are confined.

There are not many articles dealing with metallocenic iPP and mesoporous silicas and even less analyzing confinement effects. Most of them are related to the in situ polymerization topic. An approach implied the pretreatment of zirconocene with methylaluminoxane (MAO) before impregnating the catalyst [15]. Other studies showed that the amount of catalyst that can be immobilized increased by pretreating the support with MAO [16] and a superior catalytic activity was observed during polymerization. Improvements in this parameter were also achieved by substituting the microsized MCM-41 by its nanoparticles [17]. Nanocomposites of PP and MCM-41 nanoparticles achieved by in situ polymerization were described once in the literature [18] but without evaluation of properties of the resulting materials. Thus, the objective of this research is to obtain a deeper understanding on the influence of the incorporation by extrusion of either MCM-41 or SBA-15 particles to the iPP and the evaluation in the attained (nano)composites of the differences in terms of thermal stability, rheological behavior, crystalline characteristics, and confinement effects. Several techniques were used in this research, including size exclusion chromatography (SEC), scanning electron microscopy (SEM), wide and small angle X-ray Scattering (WAXD and SAXS, respectively) with synchrotron radiation, differential scanning calorimetry (DSC), thermogravimetry (TGA), and rheological experiments in the molten state.

Could polypropylene chains go inside the smaller nanometric spaces of MCM-41 helped by the shear forces applied during the extrusion process? A response to this question will contribute to the understanding of how iPP confinement affects its structure and dynamics and also its thermal degradation.

2. Materials and Methods

2.1. Materials and Chemicals

A commercially available metallocene-catalyzed isotactic polypropylene (Metocene HM562P: melt flow index of 15 g/10 min at 230 °C/2.16 kg, kindly supplied by LyondellBasell) was used in the present research as polymeric matrix. The MCM-41 and SBA-15 particles were purchased from Sigma-Aldrich (specific surface area, $S_{BET}{}^{MCM-41}$ = 966 m^2/g and $S_{BET}{}^{SBA-15}$ = 619 m^2/g; average mesopore diameter [19], $D_p{}^{MCM-41}$ = 2.9 nm and $D_p{}^{SBA-15}$ = 8.0 nm) and were used as received.

2.2. (Nano)composite and Film Preparation

Composites with different contents in particles of MCM-41 (0%, 2%, 4%, 8%, and 14% in weight) or of SBA-15 (0%, 1%, 4%, 8%, and 13% in weight) and iPP were processed by melt extrusion in a corotating twin-screw microextruder (Rondol). They were named as iPPMCM2, iPPMCM4, iPPMCM8 and iPPMCM14 for the materials prepared with MCM-41 and iPPSBA1, iPPSBA4, iPPSBA8 and iPPSBA13 for the composites obtained with SBA-15. The extruded homopolymer was labeled as iPP. Both iPP polymer and mesoporous silica (MCM-41 or SBA-15) were dried previously for 24 h under vacuum at 110 °C. A screw temperature profile of 115, 170, 180, 185 and 190 °C was used from the hopper to the die, being the length-to-diameter ratio 20:1. Then, films were obtained by compression molding at 190 °C and at 30 bar for 6 min in a hot-plate Collin (model 200 × 200) press. A relatively fast cooling process (rate around 80 °C/min) was applied between plates under pressure (30 bar) to the different films from their melt to room temperature.

2.3. Sample Characterization and Properties

The molecular weight and molecular weight distribution for the isotactic polypropylene used as polymeric matrix in this investigation were obtained by Size Exclusion Chromatography (SEC) using a Waters 150-C ALP/GPC equipped with a set of three PL-GEL MIXED-A columns from Polymer Labs. The solvent used was 1,2,4-trichlorobenzene (TCB) at 135 °C with 1 mL/min flow. The apparent molecular weight of the polymer was estimated following the standard calibration procedure using monodisperse polystyrene samples and the corresponding Mark-Howink coefficients for polypropylene [20]. The values obtained after its extrusion were 173,000 g/mol and 2.0 for polydispersity.

Experiments of high-resolution field emission scanning electron microscopy (FESEM) were carried out in a S-8000 Hitachi equipment at room temperature in different cryo-fractured sections of composites at distinct mesoporous content. Those thin sections of around 40 nm were cut by cryo-ultramicrotomy (Leica EM UC6) at −120 °C and deposited in a holder.

Thermogravimetric analysis (TGA) was performed in a Q500 equipment of TA Instruments under air or nitrogen atmosphere at a heating rate of 10 °C/min. Determination of the silica amount (MCM-41 or SBA-15) in these nanocomposites prepared by extrusion was carried out as an average of the values obtained under the two atmospheres. The resulting values of the silica content are listed in Table 1. Calorimetric analyses were carried out in a TA Instruments Q100 calorimeter connected to a cooling system and calibrated with different standards. The sample weights were around 3 mg. A temperature interval from −40 to 180 °C was studied under an inert atmosphere of nitrogen at a heating rate of 20 °C/min. For the determination of the crystallinity, a value of 160 J/g was used as the enthalpy of fusion of a perfectly crystalline material [21–23].

Real-time variable-temperature simultaneous SAXS/WAXD experiments were carried out with synchrotron radiation in beamline BL11-NCD at ALBA (Cerdanyola del Vallès, Barcelona, Spain) at a fixed wavelength of 0.1 nm. An ADSC detector was used for SAXS (off beam, at a distance of 294 cm from sample) and a Rayonix one for WAXD (at about 19 cm from sample, and a tilt angle of around 30 degrees). A Linkam Unit, connected to a cooling system of liquid nitrogen, was employed for the temperature control. The calibration of spacings was obtained by means of silver behenate and Cr_2O_3 standards. The initial 2D X-ray images were converted into 1D diffractograms, as function of the inverse scattering vector, $s = 1/d = 2 \sin \theta/\lambda$. Film samples of around $5 \times 5 \times 0.1$ mm were used in the synchrotron analysis.

Table 1. Characteristic decomposition temperatures (at a mass loss of 10 wt.%, T10%; and at the maximum variation, T^{max}) under nitrogen or air atmosphere for neat iPP homopolymer and its composites with MCM-41 or SBA-15 particles. Estimation of silica wt.% content at a specific environment and the global average.

Sample	Inert Atmosphere			Oxidative Atmosphere				Average Silica wt.% Content
	T10% (°C)	T^{max} (°C)	Silica wt.% Content	T10% (°C)	$T^{max}{}_1$ (°C)	$T^{max}{}_2$ (°C)	Silica wt.% Content	
iPP	409	463	0	233	255	-	0	0
iPPMCM2	427	464	2.3	246	287	-	2.5	2.4
iPPMCM4	437	463	3.6	244	288	-	4.7	4.2
iPPMCM8	439	462	7.7	252	299	-	8.7	8.4
iPPMCM14	441	461	13.6	260	306	336	14.5	14.1
iPPSBA1	425	472	1.4	240	275	-	1.4	1.4
iPPSBA4	439	478	3.9	242	285	-	3.9	3.9
iPPSBA8	448	479	7.7	243	284	317	7.7	7.7
iPPSBA13	452	479	12.9	249	299	310	12.7	12.8

Rheological characterization was carried out in small-amplitude oscillatory shear mode using a dynamic rotational rheometer TA Instruments ARG2 (New Castle, DE, USA). The tests were performed under nitrogen atmosphere using parallel plates of 25 mm in diameter, at a frequency range between 0.1 and 100 rad/s, and a temperature interval of 180–220 °C. All tests were carried out at small stresses in order to assure the linearity of the dynamic responses [24]. These stresses were selected performing initial stress dynamic sweeps at constant frequency of 1 rad/s.

3. Results and Discussion

Figure 1 shows the FESEM micrographs for the pristine particles of MCM-41 and SBA-15 (pictures at the top) as well as for several of the composites prepared by extrusion based on iPP and both mesoporous silicas (middle and lower pictures). Important differences are observed between the two neat silicas. Particles of MCM-41 exhibit their common irregular shape [19,25] while the ones for SBA-15 show a vermicular elongated contour [10,19] with an average size of 350 nm wide and 0.9 μm long.

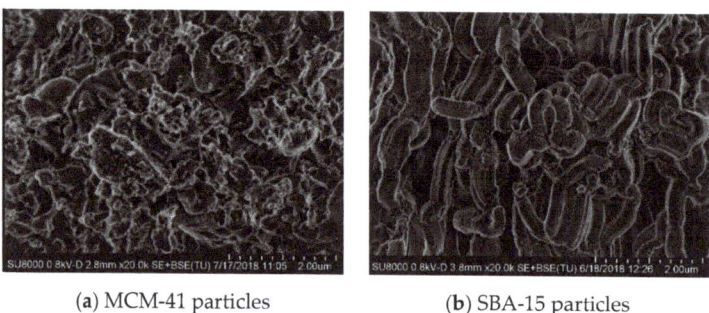

(a) MCM-41 particles   (b) SBA-15 particles

Figure 1. *Cont.*

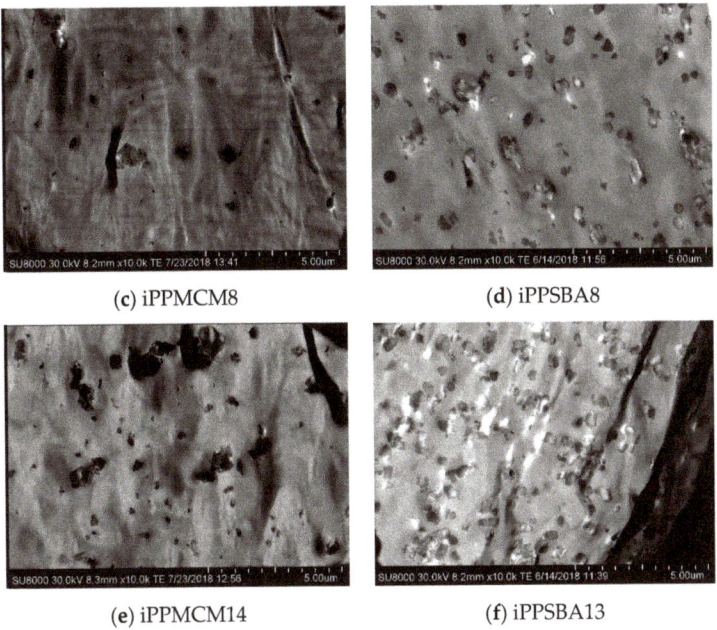

Figure 1. FESEM micrographs for different materials. On the top: pristine MCM-41 (**a**) and SBA-15 (**b**) particles; in the middle: iPPMCM8 (**c**) and iPPSBA8 (**d**) composites; and, on the bottom: iPPMCM14 (**e**) and iPPSBA13 (**f**).

Some variations are also seen when composites at almost same load are compared. A rather homogeneous dispersion of SBA-15 particles within the materials is clearly noted together with the absence of agglomerates with large size in the iPPSBA8 and iPPSBA13 composites. Moreover, an obvious increase in the number of SBA-15 particles is noticed as its content is raised in the final hybrid (precise compositions determined by TGA measurements are detailed in Table 1). Dispersion seems to be; however, less uniform in the materials containing MCM-41 microsized particles, independently of the content. In addition, MCM-41 aggregation is detected in these materials in an extent larger than in the ones prepared by using SBA-15. Despite these differences, the particle distribution and size of aggregates for both mesoporous silicas are sufficiently suitable in the final materials, mainly taking into account that they are incorporated into a non-polar polymeric matrix like the iPP, using melt extrusion as processing approach without aid of a compatibilizer agent. Existence of aggregates was also described even in composites prepared by in situ polymerization based on non-polar polyethylene and pristine microsized MCM-41 [25] or SBA-15 particles [10].

Figures 2 and 3 show the TGA curves under inert and oxidative atmosphere, respectively, for the materials prepared from iPP with either MCM-41 (plots (a)) or SBA-15 (plots (b)) particles with different silica contents. The final content in mesoporous silica, listed in Table 1, was determined from these experiments as average value of those deduced from the tests performed under these two different environments.

Figures 2 and 3 also display the degradation behavior in a broad temperature range that allows realizing the effect of the two silicas in the iPP decomposition process of these materials. First of all, incorporation by extrusion of mesoporous silica particles into iPP increases its thermal stability independently of the experimental atmosphere (Figure 2 under inert and Figure 3 under oxidant conditions, respectively) and type of silica used. Nevertheless, presence of MCM-41 or SBA-15 leads to opposite trends in the iPP degradation depending on the surrounding ambient.

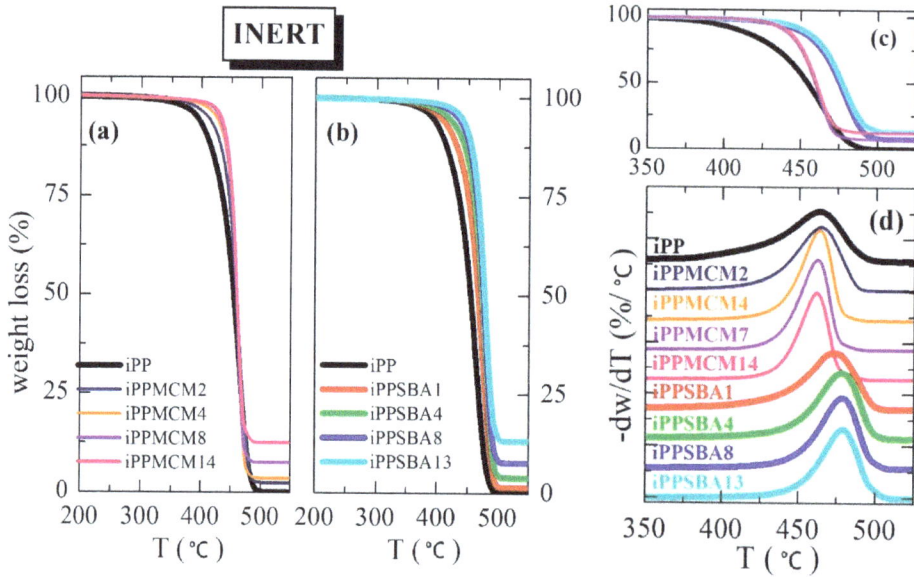

Figure 2. TGA curves under inert atmosphere for the materials prepared from iPP with particles of MCM-41 or SBA-15 ((**a**,**b**) plots, respectively) with different silica contents. Plot (**c**) shows a comparison between several samples with similar mesoporous amounts (the color code is maintained as for plots (**a**,**b**)); and plot (**d**) depicts TGA derivatives under this inert atmosphere.

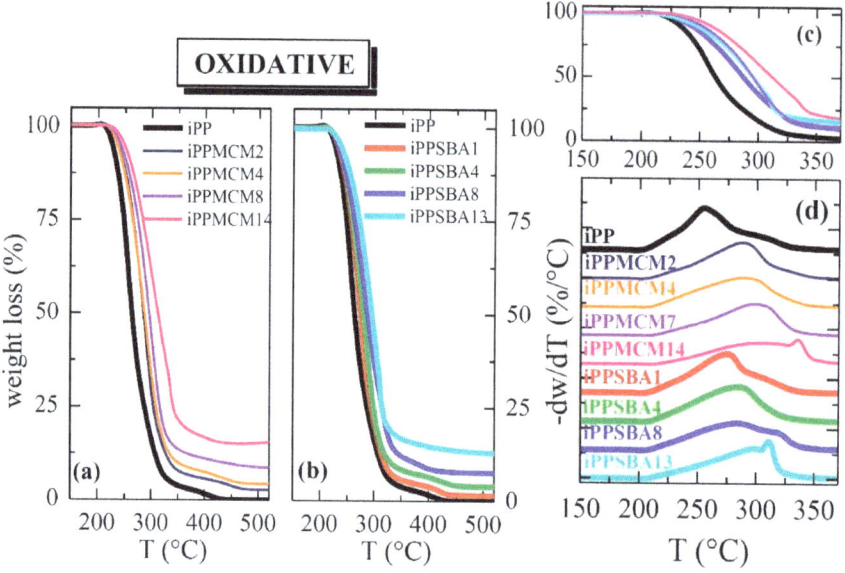

Figure 3. TGA curves under oxidant atmosphere for the materials prepared from iPP with particles of MCM-41 or SBA-15 ((**a**,**b**) plots, respectively) with different silica contents. Plot (**c**) shows a comparison between several samples with similar mesoporous amounts (the color code is maintained as for plots (**a**,**b**)); and plot (**d**) depicts TGA derivatives under this oxidative atmosphere.

Data in Table 1 together with comparison for the materials with the highest silica incorporations, depicted in the plot (c) of Figure 2, clearly show that MCM-41 stabilizes the iPP decomposition in less extent than SBA-15 particles under inert conditions. In fact, the maximum degradation temperature in these iPP–MCM-41 materials is similar or even slightly inferior to that observed in the neat iPP, as displayed in plot (d) of Figure 2. Differences are found at the beginning of the process, being these quantified in Table 1 as temperature for a mass loss of 10% by weight (T10%). Accordingly, this T10% is noticeably moved to superior temperatures in the iPP–SBA-15 composites, exhibiting values of T10% and T^{max} higher than iPP and iPP–MCM-41 hybrids, as noticed in plot (d) of Figure 2 and in Table 1. Under these conditions, all of the specimens display a single main stage of decomposition in the temperature range from 300 to 550 °C, and improvement in the iPP thermal performance associated with presence of mesoporous particles is dependent on their content. The iPP degradation mechanism was reported [26] not to change because of the SBA-15 particles although an effective delay in the build-up of the distinctive species was observed.

TGA curves of these composites under air exhibit, at least, two degradation processes in the temperature interval ranging from 200 to 375 °C, as depicted in Figure 3 (plots (a), (b) and (d)). As already described [27,28], the preliminary reaction in polyolefins during thermal oxidation is the alkyl radicals formation from polymeric chains followed by the reaction of those alkyl radicals with oxygen to form hydroperoxides, which can decompose to alkoxyl radicals. Then, the alkoxyl radicals abstract hydrogen from the chain and other alkyl radical forms. Finally, various carbonyl species are generated.

Figure 3 shows that degradation is dependent on either silica content or its pore size. The former shifts its location to higher temperatures while the latest provokes that decomposition takes place at lower temperature for materials containing SBA-15, which is the silica with significantly larger pore diameter. Consequently, a considerable thermal stabilization for the iPP matrix is achieved under oxidative conditions if MCM-41 particles are added instead of SBA-15 silica, as seen from the different plots of Figure 3 together with data listed in Table 1.

The positive impact in the iPP thermal stability in iPP–SBA-15 composites was assumed [26] to be related to an increase of the molten state viscosity in the materials by incorporation of silica. This rise was little for SBA-15 contents up to 8 wt.%, being more significant for higher SBA-15 content. Presence of SBA-15 and existence of iPP chains within their channels leaded to a hindrance of air diffusion into the bulk and, thus, to a postponement in the oxidation of iPP chains. In addition, high contents in SBA-15 provoked air diffusion through distorted pathways molten PP matrix. That assumption about the melt viscosity differences was checked by studying the rheological behavior of the different composites. Figure 4 shows the effect of both mesoporous silicas on dynamic viscosity (η') and phase angle (δ).

The iPPSBA13 material, containing the highest amount in SBA-15, presents a very considerable increase in viscosity in the whole frequency range. On the contrary, viscosity remains almost constant in the iPP–MCM-41 composites, except for the iPPMCM14 where η' is slightly raised. This different behavior could be associated with the distinct pore diameter existing within particles from these two mesoporous silicas, since both display parallel one-dimensional channels that are disposed in ordered hexagonal arrangements [29,30]. Presence of larger or smaller pores along these mesoporous particles might involve important changes, mainly related to the capability of iPP chains to be included within those bare tubes by the shear forces applied during processing. By means of evolved gas analysis, pore filling was described to be different between MCM-41 and SBA-15 in materials based on mesoporous silicas with hydroxyl-functionalized polypropylene [31], being much smaller in the former. The existence of more iPP chains within SBA-15 particles [2–4] could lead to that important variation in those two rheological parameters. These pristine iPP macromolecules coming out from the mesoporous SBA-15 could promote the matrix–filler interactions and boost an improvement in the interfacial matrix–filler adhesion, both being responsible for the rise in viscosity and δ reduction.

Truthfulness of this assumption should also involve changes of dependence of storage and loss moduli on frequency.

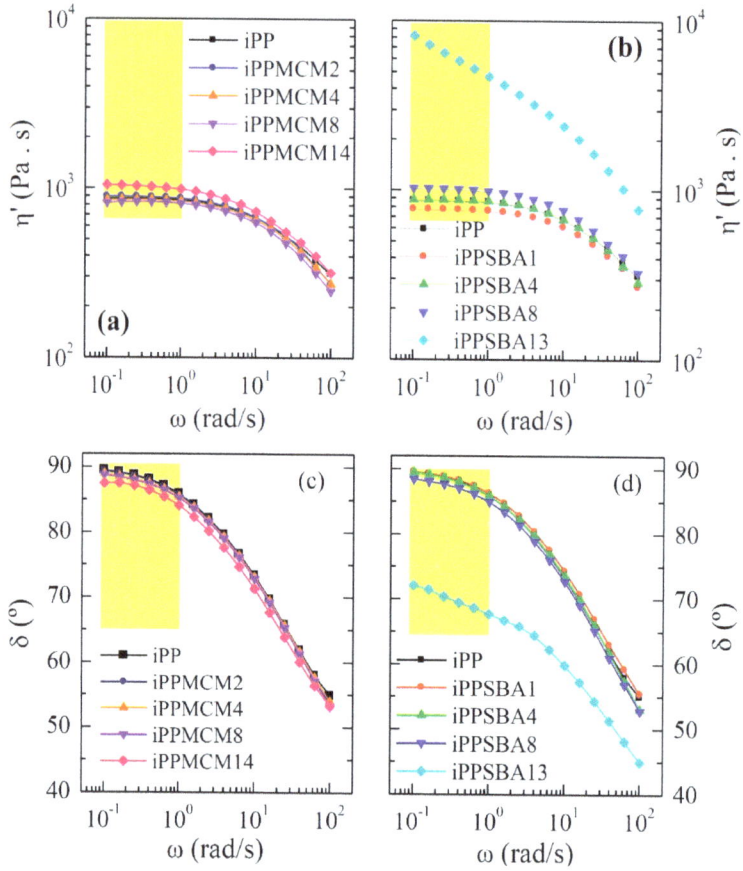

Figure 4. Reduced frequency dependence at 200 °C of viscosity, η′, (**a**,**b**) plots, and phase δ angle, (**c**,**d**) plots, for iPP–MCM-41 and iPP–SBA-15 composites at different silica contents.

Figure 5 represents the variation on frequency of elastic (G′) component of shear modulus |G*| for the different materials under study. G″ values (not shown) are higher than G′ ones in the whole frequency range. This means that terminal flow region is present in all the samples. A slight increase of both magnitudes is observed as growing amounts of mesoporous silica are loaded in the final material, either for iPP–MCM-41 or iPP–SBA-15 composites. This characteristic is more evident for the highest silica contents and in the iPP–SBA-15 hybrids. In fact, the behavior exhibited by iPPSBA13 is significantly different. Contrary to the cases of iPP homopolymer and the rest of iPP–MCM-41 or iPP–SBA-15-based materials, iPPSBA13 does not show a thermo-rheological simple response. Accordingly, the G′ and G″ isotherms do not accomplish time-temperature superposition principle in this frequency range and the corresponding master curve cannot be built. A deviation is exhibited [24] from the common dependence of G′ and G″ power-laws, ω^2 and ω respectively, as seen for G′ in Figure 5.

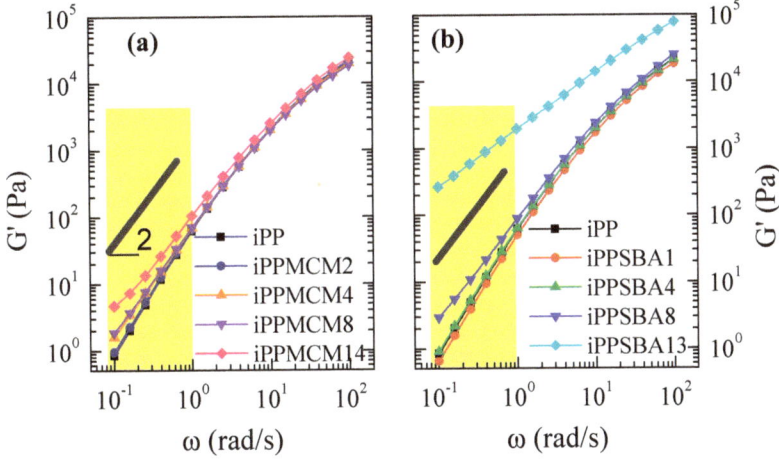

Figure 5. Reduced frequency dependence at 200 °C of storage, G'(ω), modulus for iPP–MCM-41 and iPP–SBA-15 composites at different silica contents, (**a**) and (**b**) plots, respectively. For guidance, straight lines of slope = 2 were also added.

This deviation is also clearly observed in iPPMCM14 and iPPSBA8, although both do behave as thermo-rheologically simple materials. Thus, a diminishment in slopes is attained and G' and G" values come closer at the lowest frequencies. These features point out the beginning of a transition from a liquid to a solid-like behavior. This particular rheological behavior is attributed to the development of a momentary network ascribed to percolation of fractal filler aggregates joined together with bridging polymeric macrochains [32,33]. It is designated as rheological percolation. The existence of percolation was also described in other iPP-based composites [34] although distinct trends were also found in the literature [35]. These differences could be associated with particle shape, size, state of dispersion and concentration. Each material shows its own characteristics and there are not previous articles related to iPP–mesoporous silica hybrids. Filler aspect ratio is known to be an important variable and percolation threshold is reduced with increasing ratios.

All these rheological results seem to indicate that inclusion during extrusion of iPP is into the mesoporous MCM-41 more difficult than within the SBA-15 particles, because the pore diameter is considerably smaller in the former. This difficulty would be also responsible for the analogous results found between iPPMCM14 and iPPSBA8 in spite the former contains almost double amount of silica. The fact that MCM-41 pores may be emptier than those in SBA-15 particles could also justify the aforementioned effect of MCM-41 on the oxidative iPP degradation, since those voids would favor air capture, their interaction with hydroxyl groups from silica during decomposition and the resultant reduction of air amount in the medium. This disturbance in air diffusion would contribute to postpone oxidation of iPP macrochains and to shift degradation to higher temperatures in the iPP–MCM-41 materials. However, are MCM-41 pores filled with the iPP chains or not? An absolute response cannot be provided by dependence on frequency of the rheological magnitudes in the iPP–MCM-41 composites, since they change only slightly with silica content. This fact seems to indicate that there is not a significant amount of iPP chains filling the MCM-41 channels in contrast to that observed in the iPP–SBA-15 materials.

As mentioned in the introduction, DSC measurements were previously used as an approach of easy availability for knowing the existence of polymeric chains within particles from either MCM-41 or SBA-15. The DSC curves for the present composites are displayed in Figure 6. The upper plots (a) and (b) show the first melting curves for all of specimens. They exhibit a main endotherm at about 142 °C (see results in Table 2). This process is associated with the melting of the monoclinic crystallites (as will be commented below), which are the ones commonly developed under the fast cooling conditions applied during the film processing. Total enthalpy involved in this primary process seems to remain rather unchanged with type and content of silica used in the final material. Crystallinity estimation was performed after normalization of heat flow to the actual amount of polypropylene at each specimen. Values obtained are quite similar for the distinct composites.

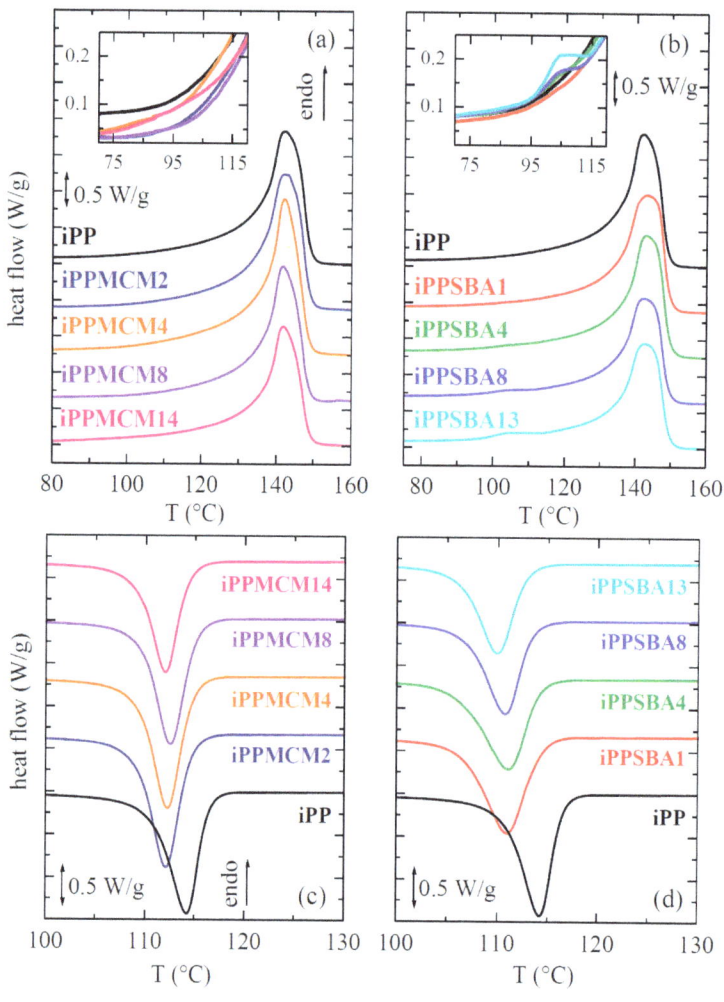

Figure 6. DSC endotherms related to the first melting run, (a,b) plots, shifted along Y axis for a better visualization, for samples prepared from iPP with MCM-41 and SBA-15 particles, respectively. DSC exotherms attained during crystallization process, (c,d) representations, for the materials extruded prepared from iPP with MCM-41 and SBA-15 silica, respectively.

Table 2. Main melting (T_m) and crystallization (T_c) temperatures; overall crystallinity (normalized to the actual iPP content in the material) estimated by DSC (f_c^{DSC}) and WAXD (f_c^{WAXD}); most probable iPP long spacing (L^{SAXS}) determined by SAXS; and crystal size (l_c) assuming a two phase model [36].

Sample	T_m (°C)	f_c^{DSC}	T_c (°C)	f_c^{WAXD}	L^{SAXS} (nm)	l_c (nm)
iPP	142.0	0.60	114.0	0.60	10.8	6.5
iPPMCM2	142.0	0.60	112.0	0.61	10.9	6.7
iPPMCM4	142.0	0.62	112.5	0.60	11.1	6.7
iPPMCM8	141.5	0.61	112.5	0.60	10.7	6.4
iPPMCM14	142.0	0.61	112.0	0.59	10.8	6.4
iPPSBA1	143.0	0.59	111.0	0.60	10.9	6.5
iPPSBA4	143.0	0.61	111.0	0.59	11.0	6.5
iPPSBA8	142.5	0.58	110.5	0.59	11.3	6.7
iPPSBA13	142.5	0.61	110.0	0.58	11.9	6.9

Another small endothermic process, at around 100 °C, is observed in the iPP–SBA-15 materials additionally to that intense melting. It was related to those small crystals that are able to be developed within the nanometric SBA-15 channels. They melt at low temperature because of their significant smaller size. On the contrary, this little endothermic event is not seen in the iPP–MCM-41 composites (the appearance of that endotherm is expected in the iPP–MCM-41 specimens at lower temperature than in those hybrids with SBA-15 particles, similarly to the features described for polyethylene materials with either MCM-41 [12–14] or SBA-15 [10,11]). Thus, it seems that there are not iPP chains within MCM-41 channels in a significant amount. Variation of rheological parameters with frequency showed, however, some slight changes with MCM-41 content. Consequently, the absolute statement of absence of iPP chains filling these nanometric MCM-41 pores cannot be established from DSC. Then, SAXS experiments turn out mandatory since they were proved as a powerful and reliable tool for the confinement analysis in iPP-based materials [2–4]. Their results will be discussed below in detail.

Lower plots, (c) and (d), in Figure 6 display the crystallization process during the subsequent cooling from the melt. Similar trend is observed in all cases, basically independent of the type and content in mesoporous silica. Accordingly, iPP crystallization is postponed in all the composites and takes place at lower temperatures. MCM-41 particles seem to delay the iPP ordering process in a smaller extent and, thus, T_c values, reported in Table 2 for the different contents, are slightly higher than those exhibited by the materials incorporating SBA-15 silica. The opposite crystallization behavior was described in HDPE/MCM-41 nanocomposites prepared by in situ polymerization using either pristine or modified with silanes MCM-41 particles. Thus, a nucleating effect was observed, which was minimized with decoration. The shift of T_c to higher temperatures was also found at the same silica range in iPP nanocomposites with SBA-15 synthesized by in situ polymerization [3]. In these latest materials, a crystallization hindrance was seen only for SBA-15 amounts greater than 20 wt.%. It should be commented that molecular weight of that synthesized iPP was considerably inferior [3] to the one of this commercial iPP here used.

Do MCM-41 and SBA-15 particles trigger the same iPP crystalline lattice in the resultant composites? An interesting polymorphic behavior is well known to take place in iPP derivatives. Thus, the iPP can crystallize into different cells by changing microstructural characteristics, crystallization parameters and other factors, such as the incorporation of specific nucleants [37–40]. Three different polymorphic modifications, α, β and γ, were described together with a phase of intermediate or mesomorphic order obtained by fast quenching [37–43]. In addition to these four modifications, a trigonal form was firstly reported in the literature [44] in 2005 for isotactic copolymers of propylene with high contents of 1-hexene [45,46] or 1-pentene [47,48], in propylene terpolymers with 1-pentene and 1-hexene [49,50] as comonomers, and in propylene terpolymers with 1-pentene and 1-heptene [51], all synthesized by using metallocene catalysts.

The iPP–SBA-15 composites prepared by extrusion showed the coexistence of orthorhombic and monoclinic modifications in specimens slowly crystallized [2] while they only exhibited the monoclinic polymorph in those fast cooled from the melt, independently of they were extruded or prepared by means of in situ polymerization [3,4]. The driving force pushing the development of those two or one crystalline lattices, respectively, was crystallization rate, rather than the presence of that mesoporous silica.

Figure 7a shows the WAXD patterns at room temperature of the different iPP–MCM-41 samples. MCM-41 silica, represented in the inset, is completely amorphous and its halo overlaps with the iPP patterns of the homopolymer and the different composites in this scattering range. No significant differences are observed between the neat iPP and the specimens containing MCM-41. All of them crystallize exclusively into monoclinic crystals, showing their characteristic diffractions. Situation changes considerably after their melting and subsequent crystallization at 20 °C/min, which is a rate much lower than that imposed during the fast cooling applied in the processing of films (approximately 80 °C/min). Now, coexistence of monoclinic and orthorhombic polymorphs is clearly seen in Figure 7b as consequence of reduction in the rate, as deduced from observation of the characteristic $(130)^{\alpha}$ and $(117)^{\gamma}$ reflections at s around 2.115 and 2.275 nm^{-1}, respectively.

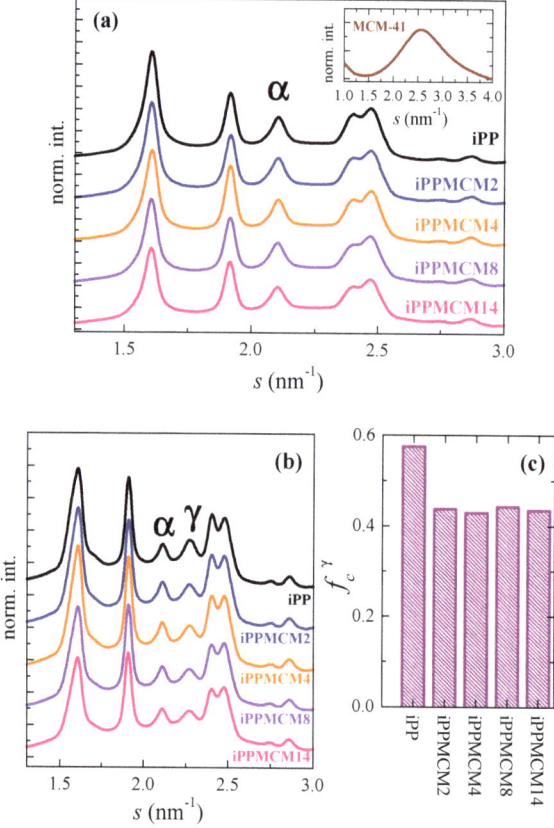

Figure 7. (a) Synchrotron WAXD 1D diffractograms of initial iPP–MCM-41 materials and the MCM-41 silica (inset); (b) WAXD 1D profiles at room temperature after crystallization at 20 °C/min from the melt; (c) orthorhombic γ content estimated from the WAXD 1D profiles at room temperature after crystallization at 20 °C/min.

Development of the γ modification is boosted in the metallocene iPPs [52,53] if its crystallization takes place slowly. Figure 7c represents the orthorhombic content deduced from those profiles represented in Figure 7b. Presence of MCM-41 within the iPP matrix hinders significantly the formation of this polymorph. In fact, its amount is reduced from 57% of the total crystals in the neat iPP to around 43% for the MCM-15 composites, being the percentage quite similar for all composites. This feature is somehow different to the results found in slowly crystallized iPP–SBA-15 composites, where the presence of SBA-15 silica reduced only very slightly the amount in orthorhombic crystals [2]. It is true that crystallization rate in those iPP–SBA-15 samples was much smaller than that applied for these iPP–MCM-41 composites.

Determination of overall crystallinity together with that for the individual content in each polymorph requires different subtractions: firstly, of the amorphous MCM-41 contribution, followed next of that coming from the amorphous iPP halo [23,43]. The values achieved after normalization to the actual iPP content in the composites are rather similar for all them, as listed in Table 2. They turn out also analogous to the crystallinity degrees estimated for the SBA-15 composites.

Another aspect to solve is to learn, without a doubt, whether or not there are iPP chains within the MCM-41 mesostructure. On one hand, the small endotherm usually observed in PE–MCM-41 [12–14], UHMWPE–SBA-15 [10,11] and iPP–SBA-15 nanocomposites [2–4], related to the melting of polymeric chains inside the pores, does not appear in the DSC curves during the first melting process of these iPP–MCM-41 materials. On the other hand, dependences on frequency of their rheological parameters seem to indicate the beginning of a transition from a liquid to a solid-like behavior, which could be ascribed to presence of iPP chains within MCM-41 channels. It must be considered that although some iPP macrochains can fill out these MCM-41 nanometric pores, the amount must be much lower than that present in the SBA-15 channels because of the significantly smaller diameter in the former. Moreover, extrusion was used for preparation of these iPP–MCM-41 composites, which can be an unfavorable approach for the inclusion of iPP within the nanometric pores compared with the in situ polymerization. As mentioned in the introduction, real-time variable-temperature SAXS measurements using synchrotron radiation were a very useful and conclusive tool to learn on the presence of iPP chains within the SBA-15 mesostructure [2–4]. Accordingly, results from these experiments are now discussed for the iPP–MCM-41 composites.

Figure 8 shows the Lorentz-corrected synchrotron SAXS 1D profiles attained from 20 to 160 °C during the initial melting at 20 °C/min in the different samples of interest. First of all, it should be commented that the vertical scale was divided for the pristine MCM-41 by a factor of 7 in order to focus the attention in the materials where silica is the minor component. Secondly, the high intensity and collimation of synchrotron radiation used proves that the first (100) order, corresponding to MCM-41 characteristic hexagonal arrangement, which is commonly asymmetric, is really split into two contributions, appearing at 0.249 nm^{-1} and 0.274 nm^{-1}, for this commercial MCM-41 silica. Superior orders (not shown) display the same feature.

Profiles for the different composites exhibit clearly two distinct regions: the one observed at the lowest s values (at around 0.085 nm^{-1} at 20 °C) and the zone located in the surrounding of 0.25 nm^{-1}, perceptible as double peak. The former is associated with the long spacing (L) ascribed to lamellar iPP crystals while the latest arises from the above commented hexagonal arrangement for the MCM-41 particles. At a first approximation, all the samples at the low s interval, except obviously the neat MCM-41 silica, show at room temperature a broad L peak of small intensity, which location is moved to smaller s values and becomes considerably more intense as increasing temperatures. No important changes seem to take place with the MCM-41 content. Regarding the double peak ascribed to the MCM-41 first order, its intensity is, obviously, significantly dependent on the amount of mesoporous silica, as noticed clearly in Figure 8.

A detailed assessment of the long spacing for the iPP component confirms a quite small effect of presence and content of MCM-41 on their values, as seen in Figure 9. The first melting of the initial samples proves that the L values range from 10.7 to 11.1 nm at room temperature (crystal size, l_c, between 6.4 and 6.7 nm, assuming a simple two phase model [36], as seen in Table 2). These are similar to those reported for the iPP–SBA-15 composites [4] with an analogous thermal treatment. Moreover, two regions are noted depending on temperature in all samples: an initial one, up to around 120 °C, with a moderate increase of L; and a final one, with a very important rise of L values, which is attributed to the crystal thickening phenomenon. This last stage coincides with the main melting endotherm (see Figure 6a) and it is ascribed to the usual melting-recrystallization processes.

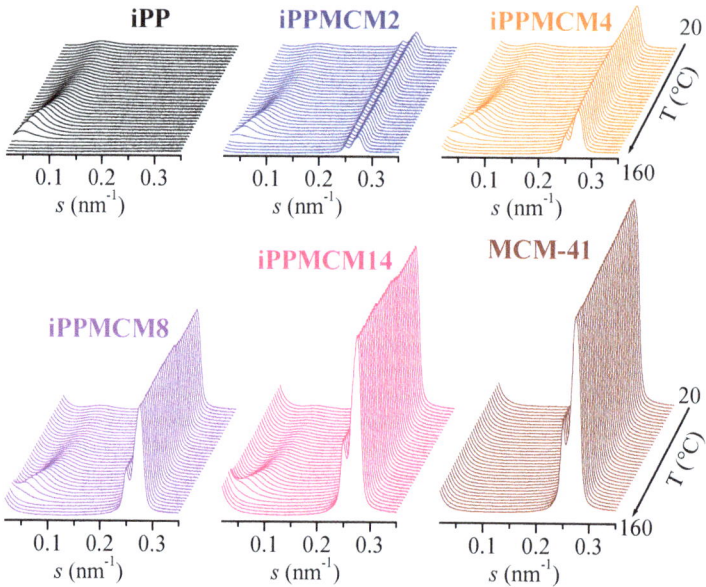

Figure 8. Lorentz-corrected synchrotron SAXS 1D diffractograms for the melting at 20 °C/min of the indicated samples. The vertical scale for neat MCM-41 was divided by a factor of 7. For clarity, only one every two frames is plotted.

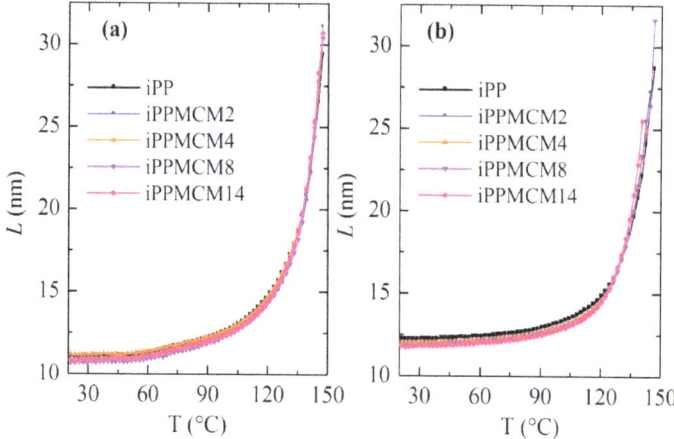

Figure 9. Dependence of Lorentz-corrected long spacing on temperature for the different iPP–MCM-41 specimens during (**a**) first melting and (**b**) melting after crystallization at 20 °C/min.

Crystallization at 20 °C/min leads to slightly thicker crystals, whose sizes at room temperature range from 11.7 to 12.1 nm, as seen in Figure 9b. It should be reminded that some of these crystallites are orthorhombic, as deduced from Figure 7b,c. These L values show a slight decrease as increasing MCM-41 content. Their dependence on temperature again displays two distinct regions, as aforementioned for the first melting. Now, melting-recrystallization processes occur in a less extent since crystallization rate was considerably smaller than that applied during processing of the films, so that now the original crystals are more perfect.

The thorough examination of the s interval at higher values shows a rather systematic behavior of the SAXS peak corresponding to the MCM-41 first order of its hexagonal arrangement in the different iPP–MCM-41 composites, which displays two well differentiated contributions, as mentioned above. Thus, although the position of these two peaks remains practically unchanged with temperature, their intensities, however, undergo a significant and regular increase for all the iPP nanocomposites in the temperature range from around 55 to 90 °C, as clearly observed in Figure 10a,b. This latest plot represents the derivative of the total area of those two peaks corresponding to the first order of that hexagonal arrangement shown by this commercial MCM-41. Interestingly, that increase is not observed for the neat MCM-41 silica (as also depicted in Figure 8), which exhibits a location, width, and intensity practically constant in the temperature interval of interest.

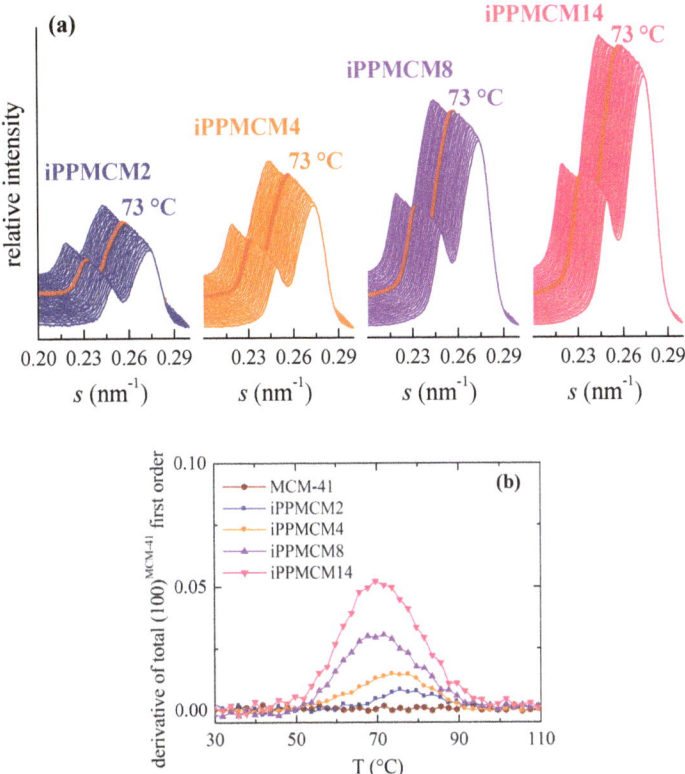

Figure 10. (a) SAXS profiles during the first melting process from 20 (rear profiles) to 160 °C (front patterns) for the distinct iPP–MCM-41composites at the s interval corresponding to the first order of MCM-41 hexagonal arrangement. Curves were shifted for a better understanding. (b) Dependence with temperature of derivative of total area from SAXS peaks for the first order in the different iPP nanocomposites and neat MCM-41 during the melting experiments.

Appearance of a discontinuity upon temperature was already described in iPP-based materials incorporating SBA-15 prepared both by in situ polymerization and by extrusion [2–4]. It was assumed that the first order of the hexagonal SBA-15 morphology was able of detecting the difference in scattering contrast between the walls and the inside of the SBA-15 channels, and this latest one is dependent on the semicrystalline or completely amorphous state of the iPP macrochains in the interior of pores due to differences in electronic density. In those iPP–SBA-15 nanocomposites, the increase of intensity took place at temperatures ranging from 95 to 120 °C and its magnitude was strongly dependent on SBA-15 content.

Change in intensity occurs now at lower temperatures, between 55 and 90 °C (as clearly depicted in Figure 10b) because the pore size in MCM-41 is much smaller than that in the SBA-15 silica: around 3.3 nm for MCM-41 and around 10 nm for SBA-15. The iPP crystals that are able to be developed within this nanometric space existing in the MCM-41 channels are, consequently, thinner than those attained in the SBA-15 pores and thus their melting takes place at inferior temperature. Intensity dependence with the MCM-41 composition is less important because the amount of iPP within the pores is much smaller in comparison with that in the SBA-15 silica. Other adverse circumstance that favors this poor pore filling is the preparation strategy since now the extrusion was used. In fact, it should be reminded that no secondary endotherm is seen in these extruded iPP–MCM-41 while a small endothermic peak was observed at same temperature interval for in situ polymerized PE–MCM-41 materials [12–14].

Accordingly, it could be expected that the DSC results are not able to discern the eventual endotherm arising from the confined crystallites considering the much lower amount of iPP chains inside the pores of MCM-41. Figure 6b indicates that in the case of SBA-15 that endotherm is clearly observed only for silica contents above around 8%. Fortunately, the SAXS measurements are sensitive enough to notice the existence of iPP chains within these nanometric channels in these iPP–MCM-41 composites, turning out a decisive technique for knowledge of the iPP confinement.

As mentioned above, the use of synchrotron radiation allows in this commercial MCM-41 silica distinguishing a regular bimodal hexagonal arrangement with primary mesopore average diameters [54] of 3.6 and 3.3 nm for the peak with the lowest and highest intensity, respectively (these two components cannot be resolved in the case of conventional X-ray radiation). Both mesopores undergo the increase in intensity that occurs when the iPP crystallites melt to its amorphous state, as depicted in Figure 10a, but, interestingly, the ratio of intensities of the two components displays a clear jump at temperatures corresponding to the melting of confined crystallites, as observed in Figure 11a. That ratio is, however, maintained constant before and after iPP crystals main melting peak (centered at around 142 °C, as deduced from Figure 6).

Moreover, pristine MCM-41 silica does not modify the ratio between the two component peaks, since nothing changes within its pores along the whole temperature interval. It follows, therefore, that the two components with significantly different pore sizes existing in the present commercial MCM-41 silica exhibit a noticeably different behavior regarding confinement of iPP crystallites. It was described through DSC experiments [14] for in situ polymerized PE–MCM-41 that crystallinity of the "secondary" endothermic process after cooling from the molten state, i.e., along the second melting process was reduced considerably in all those specimens. That observation suggested that there was a delay in the formation of those ordered entities within MCM-41 channels in those nanocomposites because of confinement effects. Crystallites could not be developed in the same extent and size during the experimental time of the DSC test. This feature involved the diminishment of the corresponding area and the shift of that secondary peak to lower temperatures. In these iPP–MCM-41 materials, this small endotherm is not detected during DSC measurements probably because the amount of iPP within MCM-41 mesostructure is rather small. Figure 11b shows also that the ratio of the two peaks from the MCM-41 first order is maintained unchanged along the whole temperature interval during the second melting. It follows that cooling from the melt at 20 °C/min allows crystallizing the iPP chains located outside mesoporous silica, as shown in Figure 6c, but not the ones filling the MCM-41 pores. The iPP chains confined within those nanometric spaces reduce significantly their crystallization kinetics and require times longer than those involved in the SAXS experiments.

All of these results show that occurrence of confinement is characterized in these iPP–MCM-41 based materials by a substantial decrease of the melting temperatures of the polymeric chains filling the pores owing to their reduced sizes and thickness of these confined crystallites. Changes during the melting processes at the s interval of the SAXS profiles where the first order peaks for the MCM-41 are observed allow assuring the existence of iPP within these nanometric pores.

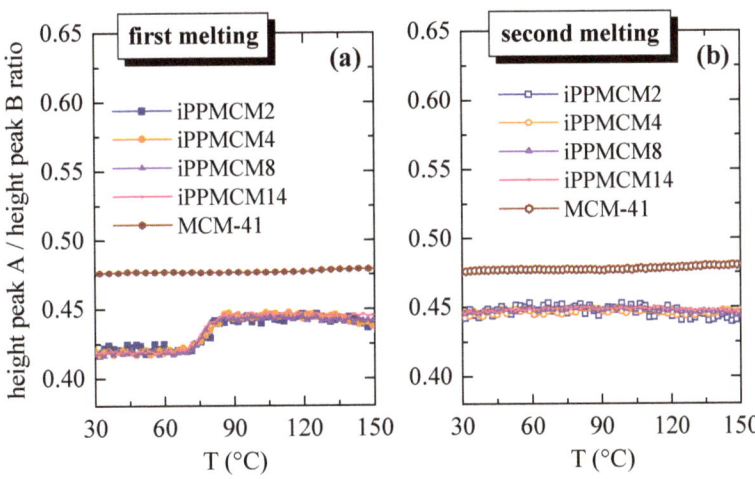

Figure 11. Variation with temperature of ratio between intensities of the two individual peaks, A and B (small and high intensity, respectively) that compose the MCM-41 first order, during: (**a**) first and (**b**) second melting processes.

4. Conclusions

Nanocomposites based on isotactic polypropylene (iPP) and mesoporous silicas, either MCM-41 or SBA-15 particles, were prepared by melt extrusion. The type of silica incorporated has affected the final characteristics found in the resulting nanocomposites. Changes were, firstly, detected in the degradation behavior, with differences were dependent on experimental environment. MCM-41 particles stabilized the iPP decomposition in a less extent than SBA-15 particles under inert conditions while they contributed to increase the thermal stability of the iPP matrix under oxidative environment.

Rheological response was also influenced by the pore size of the mesoporous silica and its content. Variations in the inclusion of iPP chains within the mesostructure of these two silicas played a key role in the iPP dynamics. Beginning of a transition from a liquid to a solid-like behavior is only intuitively observed in the iPP–MCM-41 composites while rheological percolation is clearly deduced in the iPPSBA13 material.

These differences are associated with variations in the iPP confinement within these two mesoporous silicas. Nevertheless, DSC results did not provide any information on confined macrochains in the iPP–MCM-41 materials whereas a small endotherm, attributed to the melting of the confined iPP crystallites, was clearly observed in the iPP–SBA-15 composites. Real-time variable-temperature Small Angle X-ray Scattering (SAXS) experiments with synchrotron radiation were required to undoubtedly elucidate the presence of iPP macrochains within the MCM-41 pores.

SAXS profiles showed a variation with temperature during the first melting of the total area of the MCM-41 first order SAXS peak. This clear discontinuity was centered at around 70 °C. Its location appeared, then, at significantly lower temperatures than that exhibited by the composites containing

SBA-15 particles since the MCM-41 pores are considerably much smaller than those in the SBA-15 silica. Accordingly, the iPP crystals that are able to be developed inside the MCM-41 pores in the iPP–MCM-41 nanocomposites are thinner than those attained in the materials incorporating SBA-15 and their melting takes place at inferior temperature. Furthermore, SAXS results on the present commercial MCM-41 silica indicated that it is actually constituted by two components with different average pore sizes, which exhibited a noticeably different behavior regarding confinement of iPP crystallites. Thus, the ratio of intensities of the two component peaks of the SAXS first order shows also a discontinuity at around 70 °C.

Precise information on the influence of MCM-41 on the iPP long spacing can be also deduced from these SAXS measurements since overlapping of the most probable iPP long spacing peak and the characteristic MCM-41 first-order diffraction does not occur in these materials because of their smaller pore size in this ordered hexagonal arrangement compared with that existing in the SBA-15 silica.

This study highlights the importance that pore size exerts in the confinement of iPP chains within the nanometric mesostructures of silicas with a subsequent effect in fundamental properties as thermal stability and dynamics of the resultant materials. They can contribute to spread out the already extensive application fields of iPP.

Author Contributions: Conceptualization, M.L.C.; Material Preparation, R.B.-G.; Characterization and methodology, R.B.-G., J.M.G.-E., J.A.R., L.Q., E.P., M.L.C.; formal analysis, R.B.-G., J.A.R., L.Q., E.M.V., E.P., M.L.C.; investigation, R.B.-G., J.M.G.-E., J.A.R., L.Q., E.M.V., E.P., M.L.C.; software, E.P.; writing—original draft preparation, R.B.-G., M.L.C..; writing—review and editing, R.B.-G., J.M.G.-E., J.A.R., L.Q., E.M.V., E.P., M.L.C.; project administration, M.L.C.; funding acquisition, E.P., M.L.C. All authors have read and agreed to the published version of the manuscript.

Funding: This research was funded by the Agencia Estatal de Investigación (AEI, Spain) and by the European Regional Development Fund (FEDER, UE) through the project with grant number MAT2016-79869-C2-1-P. R. Barranco-García pre-doctoral contract (BES-2014-070972) was funded by MINECO through the project with grant number MAT2013-47972-C2-1-P. Synchrotron experiments were funded by ALBA Synchrotron Light Facility.

Acknowledgments: The collaboration of the ALBA Synchrotron Light Facility personnel (especially J.C. Martínez of the beamline BL11-NCD) is grateful.

Conflicts of Interest: The authors declare no conflict of interest. The funders had no role in the design of the study; in the collection, analyses, or interpretation of data; in the writing of the manuscript, or in the decision to publish the results.

References

1. Jackson, C.L.; McKenna, G.B. The melting behavior of organic materials confined in porous solids. *J. Chem. Phys.* **1990**, *93*, 9002–9011. [CrossRef]
2. Barranco-García, R.; López-Majada, J.M.; Martínez, J.C.; Gómez-Elvira, J.M.; Pérez, E.; Cerrada, M.L. Confinement of iPP crystallites within mesoporous SBA-15 channels by Small Angle X-ray Scattering in extruded iPP-SBA-15 nanocomposites. *Microporous Mesoporous Mater.* **2018**, *272*, 209–216. [CrossRef]
3. Barranco-García, R.; Ferreira, A.E.; Ribeiro, M.R.; Lorenzo, V.; García-Peñas, A.; Gómez-Elvira, J.M.; Pérez, E.; Cerrada, M.L. Hybrid materials obtained by in situ polymerization based on polypropylene and mesoporous SBA-15 silica particles: Catalytic aspects, crystalline details and mechanical behavior. *Polymer* **2018**, *151*, 218–230. [CrossRef]
4. Barranco-García, R.; López-Majada, J.M.; Lorenzo, V.; Gómez-Elvira, J.M.; Pérez, E.; Cerrada, M.L. Confinement of iPP chains in the interior of SBA-15 mesostructure ascertained by gas transport properties in iPP-SBA-15 nanocomposites prepared by extrusion. *J. Membr. Sci.* **2019**, *569*, 137–148. [CrossRef]
5. Hammond, W.; Prouzet, E.; Mahanti, S.D.; Pinnavaia, T.J. Structure factor for the periodic walls of mesoporous molecular sieves. *Microporous Mesoporous Mater.* **1999**, *27*, 19–25. [CrossRef]
6. Sauer, J.; Marlow, F.; Schüth, F. Simulation of powder diffraction patterns of modified ordered mesoporous materials. *Phys. Chem. Chem. Phys.* **2001**, *3*, 5579–5584. [CrossRef]
7. Wang, X.; Ma, X.; Song, C.; Locke, D.R.; Siefert, S.; Winans, R.E.; Möllmer, J.; Lange, M.; Möller, A.; Gläser, R. Molecular basket sorbents polyethylenimine-SBA-15 for CO_2 capture from flue gas: Characterization and sorption properties. *Microporous Mesoporous Mater.* **2013**, *169*, 103–111. [CrossRef]

8. De Sousa, A.; de Sousa, E.M.B.; de Sousa, R.G. Synthesis and Characterization of Poly(*N*-isopropylacrylamide)/SBA-15 Silica Nanocomposites. *J. Nanosci. Nanotechnol.* **2015**, *15*, 9438–9448. [CrossRef]
9. Xu, X.; Song, C.; Andrésen, J.M.; Miller, B.G.; Scaroni, A.W. Preparation and characterization of novel CO_2 "molecular basket" adsorbents based on polymer-modified mesoporous molecular sieve MCM-41. *Microporous Mesoporous Mater.* **2003**, *62*, 29–45. [CrossRef]
10. Ferreira, A.E.; Cerrada, M.L.; Pérez, E.; Lorenzo, V.; Cramail, H.; Lourenço, J.P.; Ribeiro, M.R. UHMWPE/SBA-15 nanocomposites synthesized by in situ polymerization. *Microporous Mesoporous Mater.* **2016**, *232*, 13–25. [CrossRef]
11. Ferreira, A.E.; Cerrada, M.L.; Pérez, E.; Lorenzo, V.; Vallés, E.; Ressia, J.; Cramail, H.; Lourenço, J.P.; Ribeiro, M.R. UHMWPE/HDPE in-reactor blends, prepared by in situ polymerization: Synthetic aspects and characterization. *Express Polym. Lett.* **2017**, *11*, 344–361. [CrossRef]
12. Cerrada, M.L.; Pérez, E.; Lourenço, J.P.; Campos, J.M.; Ribeiro, M.R. Hybrid HDPE/MCM-41 nanocomposites: Crystalline structure and viscoelastic behaviour. *Microporous Mesoporous Mater.* **2010**, *130*, 215–223. [CrossRef]
13. Cerrada, M.L.; Pérez, E.; Lourenço, J.P.; Bento, A.; Ribeiro, M.R. Decorated MCM-41/polyethylene hybrids: Crystalline Details and Viscoelastic Behavior. *Polymer* **2013**, *54*, 2611–2620. [CrossRef]
14. Cerrada, M.L.; Bento, A.; Pérez, E.; Lorenzo, V.; Lourenço, J.P.; Ribeiro, M.R. Hybrid Materials Based on Polyethylene and MCM-41 Particles Functionalized with Silanes: Catalytic Aspects of In Situ Polymerization, Crystalline Features and Mechanical Properties. *Microporous Mesoporous Mater.* **2016**, *232*, 86–96. [CrossRef]
15. Marques, M.F.V.; Pombo, C.C.; Silva, R.A.; Conte, A. Binary metallocene supported catalyst for propylene polymerization. *Eur. Polym. J.* **2003**, *39*, 561–567. [CrossRef]
16. Vayá, V.I.C.; Belelli, P.G.; dos Santos, J.H.Z.; Ferreira, M.L.; Damiani, D.E. Influence of acidic support in metallocene catalysts for ethylene polymerization. *J. Catal.* **2001**, *204*, 1–10. [CrossRef]
17. Li, K.T.; Ko, F.S. Dimethylsilylbis(1-indenyl) zirconium dichloride/methylaluminoxane catalyst supported on nanosized silica for propylene polymerization. *J. Appl. Polym. Sci.* **2008**, *107*, 1387–1394. [CrossRef]
18. González, D.M.; Quijada, R.; Yazdani-Pedram, M.; Lourenço, J.P.; Ribeiro, M.R. Preparation of polypropylene-based nanocomposites using nanosized MCM-41 as support and in situ polymerization. *Polym. Int.* **2016**, *65*, 320–326. [CrossRef]
19. Watanabe, R.; Hagihara, H.; Sato, H. Structure-property relationship of polypropylene-based nanocomposites by dispersing mesoporous silica in functionalized polypropylene containing hydroxyl groups. Part 1: Toughness, stiffness and transparency. *Polym. J.* **2018**, *50*, 1057–1065. [CrossRef]
20. Brandrup, J.; Imergut, E.H.; Grulke, E.A. (Eds.) *Polymer Handbook*, 4th ed.; John Wiley and Sons: New York, NY, USA, 1999.
21. Bond, E.B.; Spruiell, J.E.; Lin, J.S. A WAXD/SAXS/DSC study on the melting behavior of Ziegler-Natta and metallocene catalyzed isotactic polypropylene. *J. Polym. Sci. Part B Polym. Phys.* **1999**, *37*, 3050–3064. [CrossRef]
22. Krache, R.; Benavente, R.; López-Majada, J.M.; Pereña, J.M.; Cerrada, M.L.; Pérez, E. Competition between alpha, beta and gamma polymorphs in a beta-nucleated metallocene isotactic polypropylene. *Macromolecules* **2007**, *40*, 6871–6878. [CrossRef]
23. Pérez, E.; Cerrada, M.L.; Benavente, R.; Gómez-Elvira, J.M. Enhancing the formation of the new trigonal polymorph in isotactic propene-1-pentene copolymers: Determination of the X-ray crystallinity. *Macromol. Res.* **2011**, *19*, 1179–1185. [CrossRef]
24. Ferry, J.D. *Viscoelastic Properties of Polymers*, 3rd ed.; John Wiley and Sons: New York, NY, USA, 1980.
25. Bento, A.; Lourenço, J.P.; Fernandes, A.; Cerrada, M.L.; Ribeiro, M.R. Functionalization of mesoporous MCM-41 (nano)particles: Preparation methodologies, role on catalytic features and dispersion within PE nanocomposites. *ChemCatChem* **2013**, *5*, 966–976. [CrossRef]
26. Barranco-García, R.; Cerrada, M.L.; Ressia, J.A.; Vallés, E.M.; García-Peñas, A.; Pérez, E.; Gómez-Elvira, J.M. Effect of mesoporous SBA-15 silica on the thermal stability of isotactic polypropylene based nanocomposites prepared by melt extrusion. *Polym. Degrad. Stabil.* **2018**, *154*, 211–221. [CrossRef]
27. Achimsky, L.; Audouin, L.; Verdu, J.; Rychly, J.; Matisova-Rychla, L. On a transition at 80 degrees C in polypropylene oxidation kinetics. *Polym. Degrad. Stabil.* **1997**, *58*, 283–289. [CrossRef]
28. Nakatani, H.; Suzuki, S.; Tanaka, T.; Terano, M. New kinetic aspects on the mechanism of thermal oxidative degradation of polypropylenes with various tacticities. *Polymer* **2005**, *46*, 12366–12371. [CrossRef]

29. Beck, J.S.; Vartuli, J.C.; Roth, W.J.; Leonowicz, M.E.; Kresge, C.T.; Schmitt, K.D.; Chu, C.T.W.; Olson, D.H.; Sheppard, E.W.; McCullen, S.B.; et al. A New Family of Mesoporous Molecular Sieves Prepared with Liquid Crystal Templates. *J. Am. Chem. Soc.* **1992**, *114*, 10834–10843. [CrossRef]
30. Zhao, D.Y.; Feng, J.L.; Huo, Q.S.; Melosh, N.; Fredrickson, G.H.; Chmelka, B.F.; Stucky, G.D. Triblock Copolymer Syntheses of Mesoporous Silica with Periodic 50 to 300 Angstrom Pores. *Science* **1998**, *279*, 548–552. [CrossRef]
31. Watanabe, R.; Hagihara, H.; Sato, H. Structure–property relationships of polypropylene-based nanocomposites obtained by dispersing mesoporous silica into hydroxyl-functionalized polypropylene. Part 2: Matrix–filler interactions and pore filling of mesoporous silica characterized by evolved gas analysis. *Polym. J.* **2018**, *50*, 1067–1077.
32. Zhang, Q.; Archer, L.A. Poly(ethylene oxide)/silica nanocomposites: Structure and rheology. *Langmuir* **2002**, *18*, 10435–10442. [CrossRef]
33. Song, Y.; Zheng, Q. Concepts and conflicts in nanoparticles reinforcement to polymers beyond hydrodynamics. *Prog. Mater. Sci.* **2016**, *84*, 1–58. [CrossRef]
34. Bahloul, W.; Bounor-Legaré, V.; David, L.; Cassagnau, P. Morphology and Viscoelasticity of PP/TiO$_2$ Nanocomposites Prepared by in Situ Sol-Gel Method. *J. Polym. Sci. Part B Polym. Phys.* **2010**, *48*, 1213–1222. [CrossRef]
35. Serrano, C.; Cerrada, M.L.; Fernández-García, M.; Ressia, J.; Vallés, E.M. Rheological and Structural Details of Biocidal iPP-TiO$_2$ Nanocomposites. *Eur. Polym. J.* **2012**, *48*, 586–596. [CrossRef]
36. Cerrada, M.L.; Benavente, R.; Pérez, E. Crystalline structure and viscoelastic behavior in composites of a metallocenic ethylene-1-octene copolymer and glass fiber. *Macromol. Chem. Phys.* **2002**, *203*, 718–726. [CrossRef]
37. Turner-Jones, A.; Aizlewood, J.M.; Beckett, D.R. Crystalline forms of isotactic polypropylene. *Die Makromol. Chem.* **1964**, *75*, 134–158. [CrossRef]
38. Brückner, S.; Meille, S.V.; Petraccone, V.; Pirozzi, B. Polymorphism in isotactic polypropylene. *Prog. Polym. Sci.* **1991**, *16*, 361–404. [CrossRef]
39. Varga, J. Supermolecular structure of isotactic polypropylene. *J. Mater. Sci.* **1992**, *27*, 2557–2579. [CrossRef]
40. Lotz, B.; Wittmann, J.C.; Lovinger, A.J. Structure and morphology of poly(propylenes): A molecular analysis. *Polymer* **1996**, *37*, 4979–4992. [CrossRef]
41. Grebowicz, J.; Lau, S.F.; Wunderlich, B. The thermal properties of polypropylene. *J. Polym. Sci. Polym. Symp.* **1984**, *71*, 19–37. [CrossRef]
42. Polo-Corpa, M.J.; Benavente, R.; Velilla, T.; Quijada, R.; Pérez, E.; Cerrada, M.L. Development of the mesomorphic phase in isotactic propene/higher a-olefin copolymers at intermediate comonomer content and its effect on properties. *Eur. Polym. J.* **2010**, *46*, 1345–1354. [CrossRef]
43. Pérez, E.; Gómez-Elvira, J.M.; Benavente, R.; Cerrada, M.L. Tailoring the formation rate of the mesophase in random propylene-co-1-pentene copolymers. *Macromolecules* **2012**, *45*, 6481–6490. [CrossRef]
44. Poon, B.; Rogunova, M.; Hiltner, A.; Baer, E.; Chum, S.P.; Galeski, A.; Piorkowska, E. Structure and properties of homogeneous copolymers of propylene and 1-hexene. *Macromolecules* **2005**, *38*, 1232–1243. [CrossRef]
45. De Rosa, C.; Dello Iacono, S.; Auriemma, F.; Ciaccia, E.; Resconi, L. Crystal structure of isotactic propylene-hexene copolymers: The trigonal form of isotactic polypropylene. *Macromolecules* **2006**, *39*, 6098–6109. [CrossRef]
46. Cerrada, M.L.; Polo-Corpa, M.J.; Benavente, R.; Pérez, E.; Velilla, T.; Quijada, R. Formation of the new trigonal polymorph in iPP-1-hexene copolymers. Competition with the mesomorphic phase. *Macromolecules* **2009**, *42*, 702–708. [CrossRef]
47. De Rosa, C.; de Ballesteros, O.R.; Auriemma, F.; Di Caprio, M.R. Crystal Structure of the Trigonal Form of Isotactic Propylene-Pentene Copolymers: An Example of the Principle of Entropy-Density Driven Phase Formation in Polymers. *Macromolecules* **2012**, *45*, 2749–2763. [CrossRef]
48. García-Peñas, A.; Gómez-Elvira, J.M.; Lorenzo, V.; Pérez, E.; Cerrada, M.L. Unprecedented dependence of stiffness parameters and crystallinity on comonomer content in rapidly cooled propylene-co-1-pentene copolymers. *Polymer* **2017**, *130*, 17–25. [CrossRef]
49. Boragno, L.; Stagnaro, P.; Forlini, F.; Azzurri, F.; Alfonso, G.C. The trigonal form of i-PP in random C3/C5/C6 terpolymers. *Polymer* **2013**, *54*, 1656–1662. [CrossRef]

50. García-Peñas, A.; Gómez-Elvira, J.M.; Pérez, E.; Cerrada, M.L. Isotactic poly(propylene-*co*-1-pentene-*co*-1-hexene) terpolymers: Synthesis, molecular characterization, and evidence of the trigonal polymorph. *J. Polym. Sci. Part A Polym. Chem.* **2013**, *51*, 3251–3259. [CrossRef]
51. García-Peñas, A.; Gómez-Elvira, J.M.; Barranco-García, R.; Pérez, E.; Cerrada, M.L. Trigonal δ form as a tool for tuning mechanical behavior in poly(propylene-co-1-pentene-co-1-heptene) terpolymers. *Polymer* **2016**, *99*, 112–121. [CrossRef]
52. Hosier, I.L.; Alamo, R.G.; Esteso, P.; Isasi, J.R.; Mandelkern, L. Formation of the α and γ Polymorphs in Random Metallocene–Propylene Copolymers. Effect of Concentration and Type of Comonomer. *Macromolecules* **2003**, *36*, 5623–5636. [CrossRef]
53. Cerrada, M.L.; Pérez, E.; Benavente, R.; Ressia, J.; Sarmoria, C.; Vallés, E.M. Gamma polymorph and branching formation as inductors of resistance to electron beam irradiation in metallocene isotactic polypropylene. *Polym. Degrad. Stabil.* **2010**, *95*, 462–469. [CrossRef]
54. Loganathan, S.; Tikmani, M.; Ghoshal, A.K. Novel Pore-Expanded MCM-41 for CO_2 Capture: Synthesis and Characterization. *Langmuir* **2013**, *29*, 3491–3499. [CrossRef] [PubMed]

© 2020 by the authors. Licensee MDPI, Basel, Switzerland. This article is an open access article distributed under the terms and conditions of the Creative Commons Attribution (CC BY) license (http://creativecommons.org/licenses/by/4.0/).

Article

PDMS Based Hybrid Sol-Gel Materials for Sensing Applications in Alkaline Environments: Synthesis and Characterization

Rui P. C. L. Sousa [1,*], Bárbara Ferreira [1], Miguel Azenha [2], Susana P. G. Costa [1], Carlos J. R. Silva [1] and Rita B. Figueira [1,*]

1. Centro de Química, Campus de Gualtar, Universidade do Minho, 4710-057 Braga, Portugal; barbarafoferreira.730@gmail.com (B.F.); spc@quimica.uminho.pt (S.P.G.C.); csilva@quimica.uminho.pt (C.J.R.S.)
2. ISISE, Departamento de Engenharia Civil, Escola de Engenharia, Campus de Azurém, Universidade do Minho, 4800-058 Guimarães, Portugal; miguel.azenha@civil.uminho.pt
* Correspondence: rui.sousa@quimica.uminho.pt (R.P.C.L.S.); rbacelarfigueira@quimica.uminho.pt or rita@figueira.pt (R.B.F.)

Received: 29 November 2019; Accepted: 16 January 2020; Published: 7 February 2020

Abstract: Nowadays, concrete degradation is a major problem in the civil engineering field. Concrete carbonation, one of the main sources of structures' degradation, causes concrete's pH to decrease; hence, enabling the necessary conditions for corrosion reinforcement. An accurate, non-destructive sensor able to monitor the pH decrease resistant to concrete conditions is envisaged by many researchers. Optical fibre sensors (OFS) are generally used for concrete applications due to their high sensitivity and resistance to external interferences. Organic-inorganic hybrid (OIH) films, for potential functionalization of OFS to be applied in concrete structures, were developed. Polydimethylsiloxane (PDMS) based sol-gel materials were synthesized by the formation of an amino alcohol precursor followed by hydrolysis and condensation. Different ratios between PDMS and (3-aminopropyl)triethoxysilane (3-APTES) were studied. The synthesized OIH films were characterized by Fourier-transformed infrared spectroscopy (FTIR), UV–Vis spectroscopy, electrochemical impedance spectroscopy (EIS) and thermogravimetric analysis (TGA). The OIH films were doped with phenolphthalein (Phph), a pH indicator, and were characterized by UV–Vis and EIS. FTIR characterization showed that the reaction between both precursors, the hydrolysis and the condensation reactions occurred successfully. UV–Vis characterization confirmed the presence of Phph embedded in the OIH matrices. Dielectric and thermal properties of the materials showed promising properties for application in contact with a high alkaline environment.

Keywords: sol-gel; PDMS; pH; hybrid; phenolphthalein

1. Introduction

The sol-gel process is a synthetic method that allows producing organic-inorganic hybrid (OIH) materials [1,2]. These materials are also known as organically modified silicates (ORMOSILs), and have a wide range of applications [3–12]. Generically, this method consists of the hydrolysis and condensation of an alkoxide to form a polymeric matrix [13–16]. The precursors and the synthesis conditions can be tuned, allowing one to obtain a product with suitable physicochemical properties according to the required application. Furthermore, OIH sol-gel materials can be doped with several species, such as corrosion inhibitors [17], electrolytes (to produce high conductivity films) [18], pharmaceutical drugs or other biomolecules [19]. These materials can also be doped with chemosensors, allowing them to obtain a polymeric matrix with sensing abilities [20–22]. A chemosensor is a molecule that is sensitive to the presence of a certain analyte and provides a detectable change in a signal, transducing a chemical signal into an action potential [23]; pH indicators are one of the main examples of this class of chemosensors, since these molecules provide a signal change with a concentration variation in H^+; pH indicators such as cresol red, bromophenol blue and fluorescein, among others, have already been successfully entrapped into a polymeric sol-gel matrix [24]. The most sol-gel precursors used in these cases are tetramethyl orthosilicate (TMOS) and tetraethyl orthosilicate (TEOS). The literature reports also entrapment of pH indicators, such as fluorescein isothiocyanate and hydroxypyrenetrisulfonic acid [25,26], in films based on polydimethylsiloxanes (PDMS). However, these studies only covered neutral pH values. PDMS is one of the most used precursors in sol-gel synthesis due to the presence of different functional groups such as amine [27], hydroxyl [28] and epoxy [29]. PDMS based precursors are commonly viscous liquids, and several applications such as coatings [30], sponges [31] or biomedical devices [32] have been reported. The different PDMS functional groups available allow the possibility to synthesize OIH gel matrices by reacting with a cross-linker, such as a silane [33–35], followed by hydrolysis and condensation. These precursors show several advantages when compared to other sol-gel precursors (e.g., TEOS or TMOS, which produce fully inorganic materials), particularly in properties such as elasticity, transparency and biocompatibility [36]. These properties allowing one to obtain OIH films suitable for optical fibre sensing applications such as the ones reported by Gao and Wang [37,38].

Concrete structures are designed to have high durability (design lifetimes typically within 50 to 100 years) with low maintenance costs. Therefore, concrete can be regarded as a stable material, with the typical internal pH at the pore network having, usually, very high values (12.5–13.5). However, due to the combination of several factors, the degradation of concrete and reinforced concrete structures (RCS) may take place due to corrosion or chemical reactions such as the alkali–silica reaction [39,40]. Moreover, the porous structure of the concrete does not provide a perfect physical barrier. This lack of imperviousness allows the progressive ingress of aggressive species at the steel/concrete interface, causing the rupture of the passivation film. The most common causes for RCS corrosion are the ingress of chloride ions, the reaction of atmospheric CO_2 with the constituents of concrete and the combination of these two processes. Carbonation of concrete occurs due to the chemical reaction of atmospheric CO_2 with the alkaline components present in the concrete pore solution, forming calcite ($CaCO_3$). This reaction leads to a decrease in the pH values of the interstitial concrete pore solution [41]. This pH decrease, generally from values above 12.5 to values between 9 and 6, compromises the concrete's protective function [42]. Therefore, monitoring concrete and RCS is a crucial activity, not only economically but also for human safety; pH monitoring in concrete has been approached by the scientific community already. For instance, Behnood et al. [43] reviewed and compared the methods for monitoring pH in concrete and divided them into destructive and non-destructive methods. Destructive methods are the most used for concrete condition assessment. However, these are limited by sampling, which may not be representative of the whole structure [44]. Behnood et al. described the three most used destructive methods for pH monitoring; namely: the pore water expression method, that extracts the concrete pore solution under hydraulic pressure; the in-situ leaching method, which analyses the concrete pore solution inside a cavity by equilibrium with an added solution; and the ex-situ leaching method, that is based on a powder solution and by equilibration with an added solution. Pore water expression is the most used from the three methods above, and it was recommended by Plusquelle et al. [45] mainly due to its simplicity. On the other hand, non-destructive methods, which are becoming more and more evolved, can be electrochemical or optical [44]. These monitoring methods started with the development of electrochemical sensors for the measurement of corrosion potential of rebars [44]. However, in the last few years, there has been a huge interest in the development of new optical sensors which are generally based on properties such as fluorescence, absorbance, reflectance and refractive index [46]. Optical fibre sensors (OFS) for applications in concrete structures show several advantages when compared to electrochemical methods, such as reduced cost, size and weight, as well as higher sensitivity [39]. Behnood et al. concluded that OFS can be very effective for real-time pH monitoring. Nevertheless, the sensors already reported for application in concrete need to be further developed in order to obtain accurate levels of resolution, repeatability and reproducibility [43].

The use of OFS for concrete monitoring was introduced long ago, mainly for crack monitoring [47] and other mechanical variations. Since then, OFS based on sol-gel materials have been developed for structural health monitoring [39,43,45,48–50], including parameters such as temperature, humidity and pH [39]. However, most of the developed OFS functionalized with OIH sol-gel films have limitations for concrete applications, such as the leaching of the doped species from the films or poor resistance to be used in fresh concrete. As far as the authors know, no OFS functionalized with PDMS-based OIH sol-gel films for concrete pH monitoring have been described. PDMS hydroxyl-terminated has been reported as a concrete additive [51] to reduce the production of calcite protecting the concrete structures against the reduction of the pH level. Sidek et al. [52,53] reported the preparation of a fibre Bragg grating (FBG) sensor for strain detection in concrete structures, with and without a PDMS coating. PDMS coated FBG sensors have shown themselves to be more sensitive than uncoated FBG sensors [54]. Later, in 2016, Tan et al. [55] showed that PDMS-coated FBG indeed enhanced the sensibility of strain detection and that the response of the sensor was linear to the rebar corrosion rate, estimated by the weight loss. PDMS has also some limitations, such as its low Young's modulus and the possibility to absorb impurities during the curing process [55]. Nevertheless, the combination of the OFS technology and PDMS-based OIH show the potential to build a non-destructive monitoring system that can withstand the harsh conditions of fresh concrete. Phenolphthalein (Phph) is one of the most well-known pH indicators and has its turning point at a pH between 8 and 10. This indicator is already used for qualitative assessment of the depth of carbonation in concrete (e.g., by staining concrete cores right after extraction from a given structure) [56]. Therefore, OIH sol-gel materials doped with Phph can be used to functionalise OFS, allowing one to assess the carbonation of concrete structures.

This work reports a study that is only focused on the synthesis and characterization of three OIH sol-gel materials based on PDMS-diglycidyl, ether terminated (PDMS(800)-GET). The chemical characterization and stability testing of the OIH matrices in contact with simulating concrete pore solution (SCPS) were carried out. The development of OFS based on OIH sol-gel materials opens up the possibility of producing highly accurate and reliable sensing systems. Therefore, the aim of these studies was to assess the ability of these materials to be used as sensing material for OFS functionalisation. The produced films, doped and undoped with Phph, were characterized by Fourier-transformed infrared (FTIR) spectroscopy, UV-Vis spectroscopy, electrochemical impedance spectroscopy (EIS) and thermogravimetric analysis (TGA). UV–Vis characterization confirmed the presence of Phph embedded in the OIH matrices. The dielectric and thermal properties of the materials showed promising properties for application in contact with a high alkaline environment.

2. Materials and Methods

2.1. Materials

Poly(dimethylsiloxane)-diglycidyl ether (PDMS(800)-GET, Sigma-Aldrich, St. Louis, MO, USA), 3-aminopropyltriethoxysilane (3-APTES, 99%, Acros Organics, Geel, Belgium), phenolphthalein (Merck, Darmstadt, Germany), absolute ethanol (EtOH, PanReac, Darmstadt, Germany), citric acid monohydrate (Merck, Darmstadt, Germany), potassium bromide (KBr, FTIR grade, ≥99%, Sigma-Aldrich, St. Louis, MO, USA), calcium hydroxide ($Ca(OH)_2$, 95%, Riedel, Bucharest, Romania) and potassium hydroxide (KOH, 90%, PanReac, Darmstadt, Germany) were used as received. High purity deionized water with high resistivity (higher than 18 MΩ cm) obtained from a Millipore water purification system was used in all prepared solutions.

2.2. Synthesis of Organic-Inorganic Hybrid (OIH) Films

2.2.1. General Procedure

The first stage of the hybrid films' synthesis consisted of the formation of a covalent bond between PDMS(800)-GET and 3-APTES (Table 1, Figure 1). Different ratios between the two precursors were used (1:2, 1:2.5 and 1:5 PDMS: 3-APTES). The two precursors were mixed in a glass container for 20 min. A 0.22 M citric acid ethanolic solution was added to set a ratio of 0.094 between citric acid and the corresponding amount of silane. The mixture was stirred for another 20 min and water was added to set a H_2O/PDMS ratio of 29.7. The solution was stirred for more 20 min to obtain a homogeneous mixture and cast into a Teflon® mould. The Teflon® mould was sealed with parafilm® and placed in a universal oven (UNB 200, Memmert, Buechenbach, Germany) and kept at 40 °C for 15 days in order to ensure the curing of the films and evaporation of the remaining solvents. Films were identified as AES(800)-1/2, AES(800)-1/2.5 and AES(800)-1/5.

Table 1. Adopted codes for organic-inorganic hybrid (OIH) films.

OIH Films	PDMS:APTES Ratio	Phph:PDMS Ratio
AES(800)-1/2	1:2	-
AES(800)-1/2.5	1:2.5	-
AES(800)-1/5	1:5	-
AES(800)-1/2-Phph-0.01	1:2	1:100
AES(800)-1/2-Phph-0.02	1:2	1:50
AES(800)-1/2-Phph-0.05	1:2	1:20

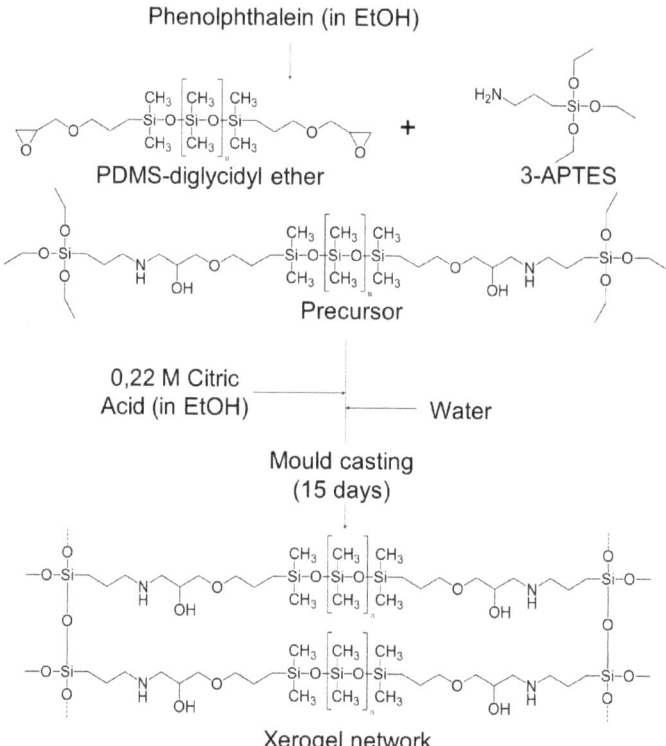

Figure 1. Schematic of the main steps used in the synthesis of PDMS based OIH films.

2.2.2. Phph Doped Films

PDMS(800)-GET (1 mmol) was stirred with a 0.1 M ethanolic solution of Phph, according to Table 1, for 20 min in a glass container; 3-APTES (2 mmol, 1/2 PDMS/3-APTES ratio) was added to the solution and the general procedure was followed. However, in this particular case water was not added into the solution. Films were identified as AES(800)-1/2-Phph-0.01, AES(800)-1/2-Phph-0.02 and AES(800)-1/2-Phph-0.05.

2.3. Characterization of the OIH Films

The OIH pure films were characterized by Fourier-transformed infrared (FTIR) spectroscopy, UV-Vis spectroscopy, electrochemical impedance spectroscopy (EIS) and thermogravimetric analysis (TGA). The three OIH films doped with Phph were only characterized by UV-Vis and EIS.

2.3.1. FTIR

FTIR spectra for AES(800)-1/2 were recorded in transmittance mode on a Bomem MB104 spectrometer, by averaging 20 scans at a maximum resolution of 4 cm^{-1}. Spectra were obtained in 4000–700 cm^{-1} range on KBr pellets. For the analysis during synthesis, a 2 µL droplet of the solution was placed in the previously prepared 100 mg KBr pellet. Droplets were collected 20 min after both precursors were added, before hydrolysis and condensation of amino alcohol precursor. Liquid precursors KBr pellets followed the same preparation. For the final OIH films, KBr pellets were prepared with 0.1 mg of cured material and 200 mg of KBr.

2.3.2. UV–Vis Spectroscopy

UV–Vis spectra for both pure and doped OIH films were recorded in absorbance mode on a Shimadzu UV-2501 PC spectrophotometer. Spectra were obtained in the range of 200–800 nm for solid samples.

2.3.3. Electrochemical Impedance Spectroscopy (EIS)

EIS measurements were carried out on all the produced OIH disc materials at room temperature in a Faraday cage, using a potentiostat/galvanostat/ZRA (Reference 600+, Gamry Instruments, Warminster, PA, USA). EIS measurements were used to characterize resistance, electrical conductivity and electric permittivity of the prepared OIH disc materials. The capacitance of OIH films was also determined. Measurements were performed using a support cell as reported in previous studies [57]. The disc films were placed between two parallel Au electrodes (10 mm diameter and 250 µm thickness) using the mentioned support cell and the EIS measurements were conducted [57]. EIS studies were accomplished by applying a 10 mV (peak-to-peak, sinusoidal) electrical potential within a frequency range from 1×10^6 Hz to 0.01 Hz (10 points per decade) at open circuit potential (OCP). The frequency response data of the studied electrochemical cells were displayed in a Nyquist plot, using Gamry ESA410 Data Acquisition software that was also used for data fitting purposes. For chemical stability studies in an alkaline environment, the OIH discs were immersed in a simulative concrete pore solution (SCPS) with a pH of 13.5. The SCPS was prepared, using deionized water at room temperature according to M. Sanchez et al. [58] and F. J. Recio et al. [59]. Generically, the SCPS was obtained by addition of 0.2 M KOH to a $Ca(OH)_2$ saturated solution [58,59].

2.3.4. TGA

Thermal analysis was carried out on an SDT Q600 system for the three pure OIH materials. Samples were subjected to a temperature ramp of 10 °C/min between room temperature and 800 °C at a constant 40 mL/min nitrogen flux. For each analysis, 15 mg of each OIH material was placed into an alumina pan.

3. Results and Discussion

3.1. FTIR Analysis

FTIR analysis was carried out for AES(800)-1/2 and is shown in Figure 2. Spectra of both precursors ((a) and (b)), from a sample during synthesis after 20 min of precursors addition (e.g., before hydrolysis and condensation of amino alcohol precursor) (c) and the final OIH material (d) are depicted.

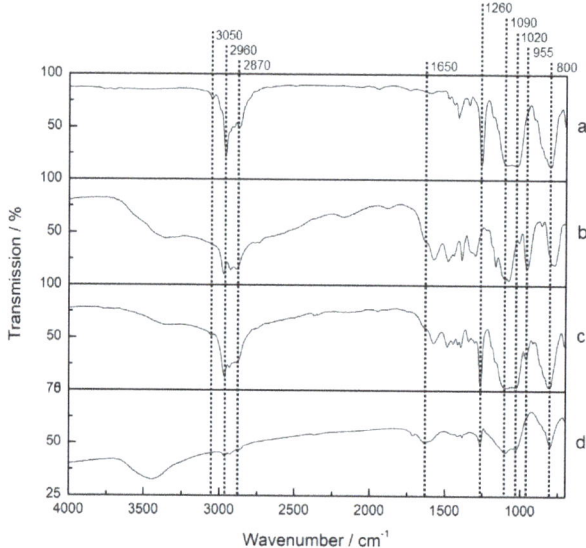

Figure 2. FTIR transmittance spectra of AES(800)-1/2. (**a**) PDMS(800)-GET; (**b**) 3-APTES; (**c**) AES(800)-1/2 during synthesis; (**d**) AES(800)-1/2 final hybrid (presented with a different scale, for higher definition purposes).

PDMS(800)-GET spectrum (a) from Figure 2 shows a weak band at 3050 cm^{-1}, corresponding to the epoxy terminal group [60]. Bands between 2960 and 2870 cm^{-1} may be ascribed to C-CH$_2$ asymmetric and symmetric stretching vibrations. A well-defined sharp peak at 1260 cm^{-1} belonging to C–Si bond symmetric bending is especially characteristic of dimethylsiloxane chains [61]. Broad bands at 1090 and 1020 cm^{-1} are related to Si–O–C and Si–O–Si bonds and the peak at 800 cm^{-1} corresponds to C–Si bond symmetric rocking vibration [61,62]. 3-APTES spectrum (b) from Figure 2. also shows the bands corresponding to C–CH$_2$ asymmetric and symmetric stretching vibrations. Besides, a broad band between 3500–3300 cm^{-1} may be ascribed to N–H bond stretching vibrations of the terminal amine group involved in H-bonds. Peaks at 1100, 1080 and 955 cm^{-1} are characteristic of Si–O–CH$_2$CH$_3$ bonds of this precursor [61]. In the spectrum of the hybrid film during its synthesis (c) from Figure 2, the epoxy group peak present in PDMS spectra (3050 cm^{-1}) disappeared, as expected, due to the reaction between this group and the amine group from the silane. The broad band between 3500–3300 cm^{-1} is also present, showing no difference between primary (3-APTES) and secondary (hybrid films) amines. The peak corresponding to C–Si symmetric bending (1260 cm^{-1}) is still present, as well as the peaks between 2960 and 2870 cm^{-1} belonging to C–CH$_2$ asymmetric and symmetric stretching vibrations and the peaks at 1090, 1020 and 800 cm^{-1}, from the dimethylsiloxane chain. The Si–O–CH$_2$CH$_3$ characteristic peak (955 cm^{-1}) is still present but on a minor scale. Final hybrid film spectrum Figure 2. (d) does not show this peak, due to the total hydrolysis of these alkoxy groups. A small peak at 1650 cm^{-1} appears, which may belong to the C–NH–C bond bending vibration, corresponding to the formed amino alcohol bond. The broad and intense band at about 3450 cm^{-1} can be related not only to the hydroxyl group formed in the film but also to water molecules that remained entrapped in the polymeric matrix.

3.2. UV–Vis Analysis

Optical absorption spectra for the three pure OIH samples are shown in Figure 3. Even though the samples are not transparent, absorbance in the range of 400–800 nm is constant, showing the films can be doped with a colorimetric chemosensor that absorbs in this range.

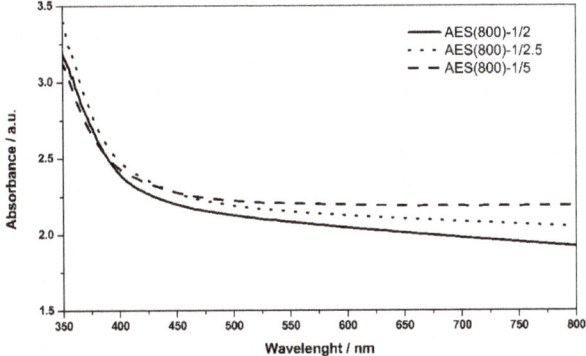

Figure 3. Absorption spectra of pure hybrid samples.

Phph, a well-known pH indicator, colourless below pH 8 and pink above pH 10, was immobilized in the sol-gel matrix. UV–Vis spectroscopy analysis was carried out to confirm the presence of the indicator in the matrix. The spectra for AES based Phph doped films are shown in Figure 4. The absorption spectrum of the basic form of Phph in ethanol is also shown as inset. The three spectra show a peak around 566 nm, corresponding to the basic form of Phph, confirming its presence within the OIH material. Since the Phph peak in ethanol and within the matrix share the same wavelength, it can be assumed that the molecule does not change the electron distribution profile that is involved in the radiation absorption.

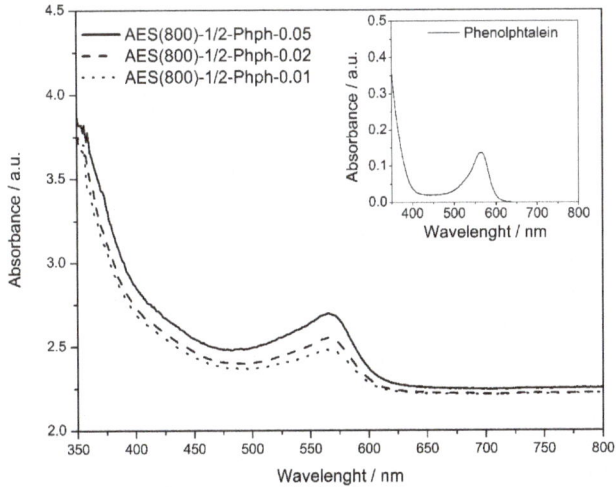

Figure 4. Absorption spectra of AES based FFT doped films. Inset: absorption spectrum of the basic form of Phph in ethanol.

The results show (Table 2) that the absorption coefficients for Phph on the three doped films are of the same magnitude order. That result was expected, since the matrix has a similar composition. Distinctively from the ethanol solution, the molecule entrapped within a rigid matrix contributes to molecule–matrix interaction, reducing the efficiency of the radiation absorption.

Table 2. UV–Vis data values obtained from the doped OIH materials spectra (Figure 4).

Phph	Ethanol	AES(800)-1/2-Phph-0.01	AES(800)-1/2-Phph-0.02	AES(800)-1/2-Phph-0.05
Absorbance (566 nm) *	0.136	0.411	0.482	0.627
[Phph]/mol g^{-1}	1.27 × 10^{-7}	8.05 × 10^{-6}	1.48 × 10^{-5}	2.62 × 10^{-5}
Optical path/cm	1.000	1.131 **	0.919 **	1.017 **
Absorption coefficient/10^4 g mol^{-1} cm^{-1}	107.08	4.52	3.55	2.35

* For the OIH films, the absorbance of the corresponding OIH undoped (AES(800)-1/2) was subtracted. ** Thickness of the film.

3.3. EIS Analysis

EIS is a technique widely used for materials characterization. EIS allows one to obtain the dielectric properties of the OIH films and can be extremely helpful on optimizing the OIH materials in the synthesis procedure. Considering the information that can be extracted (e.g., capacitance, resistivity, dielectric permittivity, etc.) from EIS measurements, it is possible to follow the degradation of the OIH materials by quantifying the changes in the dielectric properties [63,64]. When EIS is performed on high performance materials, the Nyquist plots show a typical capacitive response over a wide range of frequencies. In the high-frequency region of the Nyquist plots, the dielectric properties of the material can be extracted. In this work, the EIS was used to characterize the dielectric properties (e.g., conductivity, capacitance and electric permittivity), which can be linked to the degradation properties of the material. Therefore, these studies allow one to assess the potential of the OIH matrices to be used in contact with a high alkaline environment (fresh concrete). Figures 5–7 show the Nyquist plots of the pure OIH materials identified according to Table 1. Figure 8 shows the Nyquist plots of the OIH materials doped with Phph, which were also identified according to Table 1. The fitting results are also displayed in all the Nyquist plots. The equivalent electrical circuit (EEC) to describe the impedance spectra response was also included in each figure.

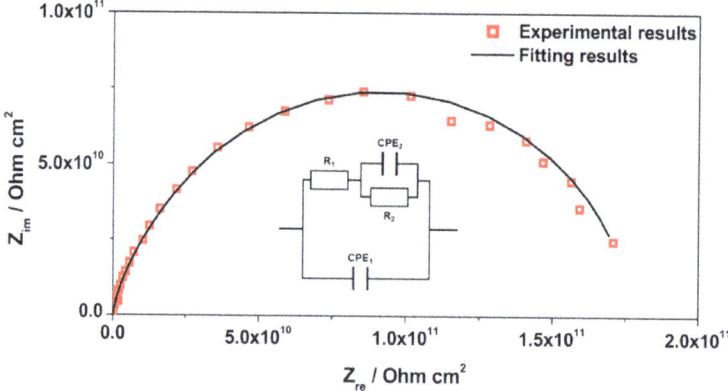

Figure 5. Nyquist plot for pure OIH AES(800)-1/2.

Figures 5–7 show that at higher frequencies both Nyquist plots describe a semicircle, with the exception of the AES(800)-1/2.5 sample (Figure 6), which intersects the x-axis; none of the other samples displayed this behaviour. The amplitudes of the different samples change with their composition, and the undoped samples show the higher impedance magnitudes. The AES(800)-1/2 sample (Figure 5) showed the highest magnitude, and was followed by the AES(800)-1/2.5 (Figure 6). Sample AES(800)-1/5 showed the lowest magnitude. These differences are assigned to the dielectric properties of the OIH materials (e.g., resistivity and capacitance). The data obtained at lower frequencies describes a line in AES(800)-1/5 (Figure 5). The same behaviour was found in all OIH materials doped with Phph. This suggests that another electrochemical process, which is attributed to the Au|OIH material interface, is taking place. This part of the Nyquist plot describes a simple parallel resistance and capacitor EEC; however, the fitting was disregarded, since it is not relevant for the comprehension of the dielectric properties of the OIH materials. The analysis of EIS data was based on the represented EEC. Constant phase elements (CPE) were used instead of pure capacitance, since the results obtained (Figures 5–8) do not show ideal behaviour. The impedance of a CPE can be defined as [65]:

$$Z_{CPE} = 1/[Q(j\omega)^\alpha]. \tag{1}$$

When $\alpha = 1$, Q represents the capacity, and if $\alpha \neq 1$, the system shows a behaviour that is linked to the surface heterogeneity. If the system shows this second behaviour instead of an ideal capacitor, the impedance for the EEC is determined by [66]:

$$Z_{CPE} = R_{Sample}/[1 + (j\omega)^\alpha QR_{Sample}], \tag{2}$$

where R_{Sample} is the resistance in parallel with the CPE. The CPE parameter Q cannot be considered as the interfacial capacitance (C_{eff}). C_{eff} is determined by [66,67]:

$$C_{eff} = [QR_{sample}^{(1-\alpha)}]^{1/\alpha}. \tag{3}$$

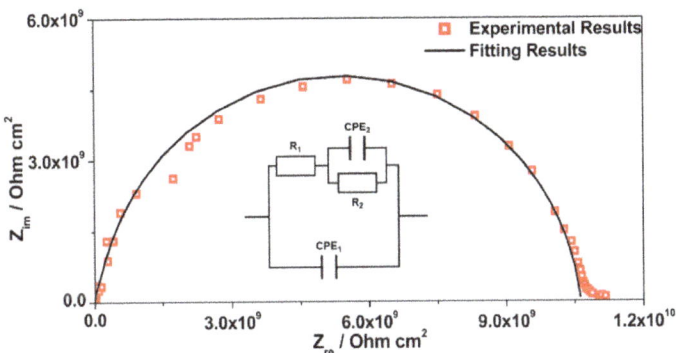

Figure 6. Nyquist plot for pure OIH AES(800)-1/2.5.

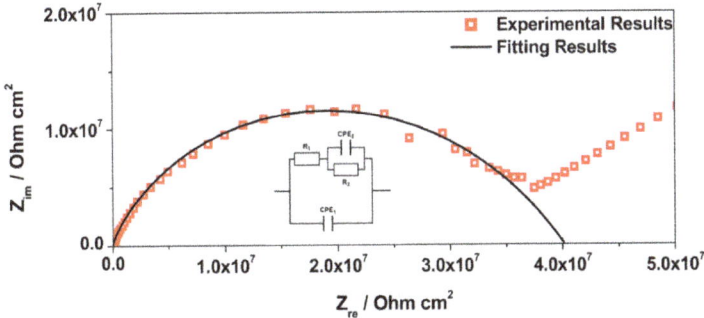

Figure 7. Nyquist plot for pure OIH AES(800)-1/5.

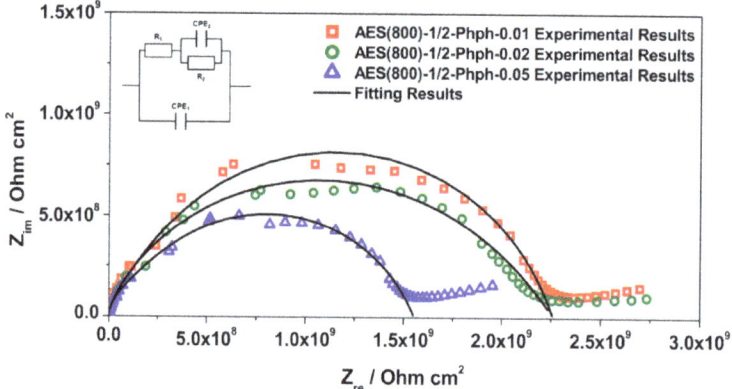

Figure 8. Nyquist plot for Phph doped OIH AES(800)-1/2-Phph films.

The EEC used to fit the Nyquist plots obtained for all samples contains two CPEs (CPE$_1$ and CPE$_2$) and two resistances (R$_1$ and R$_2$) which are related to the resistance of the OIH material. The observation of the Nyquist plots (Figures 5–8) shows, at higher frequencies, two partially overlapping semicircles with different radii. This behaviour may be identified as two time-dependent charge relaxation processes with two different time-constants (CPE$_1$ and CPE$_2$). The systems show a frequency dispersion linked to the relaxation phenomena that may be explained by the presence of residual solvent incorporated within the OIH matrix. The elements R$_1$ and R$_2$ depict the sample's (bulk) resistance of these two distinct dielectric media and according to this EEC, the sample's bulk resistance is the sum of R$_1$ and R$_2$. Three measurements were performed for each sample. However, only representative values together with the respective fitting parameters and the percentage of error associated with each element parameter are shown in Table 2.

Table 3 shows that all the resistances are within the range of 10^7–10^{11} Ω cm^2. This suggests that the studied OIH materials display suitable resistance values which may imply that the synthesized materials can endeavour the harsh conditions of the fresh concrete [68]. The electrical resistance of the OIH AES(800)-1/2 is the highest, and the resistance of the OIH AES(800)-1/5 is the lowest (Table 2). This is in accordance with the literature [7,68], since the higher the ratio between the organic and inorganic amounts of precursors, the lowest the resistance. Table 2 also shows that the OIH materials doped with Phph are of the same magnitude order (10^9). Moreover, the samples AES(800)-1/2-Phph-0.01 and AES-1/2Phph-0.02 have equal electrical resistance values. This behaviour is expected, since the ratio between PDMS and 3-APTES is the same for the OIH materials doped with Phph.

To assess the chemical stability and resistance of the pure OIH materials, these were immersed in SCPS, which was prepared as described in Section 2.3.3. This solution was used since it is generally accepted as the most representative of concrete pore solution [58,59], because it simulates the alkalinity and the high alkali content of the pore solution existent in the concrete structures. The EIS measurements of the OIH films in contact with SCPS were conducted after 30 min, 4 h and 24 h of immersion. The Bode plots obtained are shown in Figures 9–11.

Table 3. EEC data parameters values obtained by fitting the EIS response obtained for both doped and undoped films (Figures 5–8).

Sample	$CPE_1/S^\alpha\,\Omega^{-1}\,cm^{-2}$	α_1	$R_1/\Omega\,cm^2$	$CPE_2/S^\alpha\,\Omega^{-1}\,cm^{-2}$	α_2	$R_2/\Omega\,cm^2$	$R_{sample}/\Omega\,cm^2$	χ^2
AES(800)-1/2	9.27×10^{-12} (1.54%)	0.900 (0.25%)	9.00×10^{10} (0.01%)	8.12×10^{-12} (2.75%)	0.900 (0.51%)	8.84×10^{10} (2.20%)	1.78×10^{11}	2.24×10^{-1}
AES(800)-1/2.5	6.45×10^{-12} (2.17%)	0.975 (0.19%)	1.47×10^{9} (13.10%)	5.35×10^{-12} (8.36%)	0.881 (2.90%)	9.16×10^{9} (2.61%)	1.06×10^{10}	4.81×10^{-3}
AES(800)-1/5	2.46×10^{-11} (±13.19%)	0.910 (±0.97%)	2.10×10^{6} (±9.90%)	5.28×10^{-10} (±10.64%)	0.575 (±2.57%)	3.81×10^{7} (±3.00%)	4.02×10^{7}	3.58×10^{-4}
AES(800)-1/2-Phph-0.01	6.60×10^{-12} (±2.73%)	0.961 (±0.23%)	3.17×10^{8} (±7.96%)	2.97×10^{-11} (±7.11%)	0.748 (±2.56%)	1.94×10^{9} (±2.24%)	2.26×10^{9}	2.61×10^{-3}
AES(800)-1/2-Phph-0.02	7.61×10^{-12} (±3.26%)	0.964 (±0.27%)	1.94×10^{8} (±15.50%)	6.47×10^{-10} (±5.95%)	0.597 (±3.01%)	2.07×10^{9} (±2.63%)	2.26×10^{9}	1.88×10^{-3}
AES(800)-1/2-Phph-0.05	8.32×10^{-12} (±2.61%)	0.954 (±0.00%)	2.82×10^{8} (±7.40%)	5.99×10^{-11} (±7.96%)	0.713 (±3.34%)	1.27×10^{9} (±3.01%)	1.55×10^{9}	1.31×10^{-3}

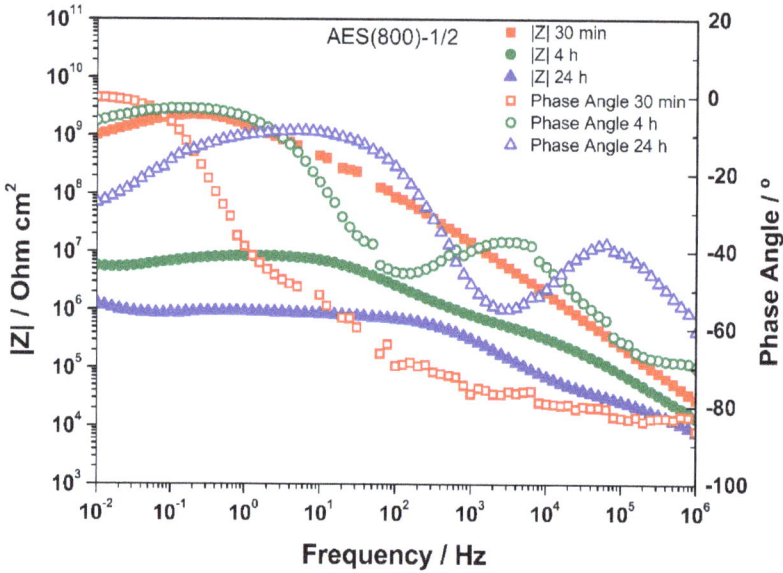

Figure 9. Bode plot for EIS analysis of AES(800)-1/2 after 30 min, 4 h and 24 h in SCPS.

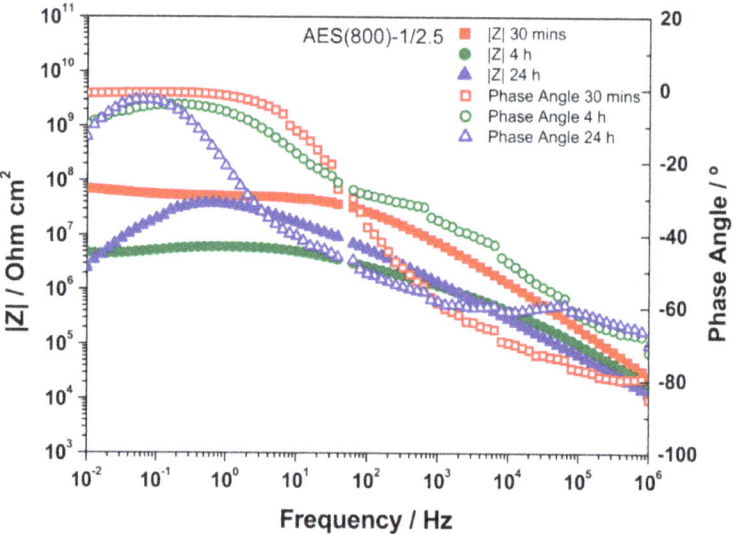

Figure 10. Bode plot for EIS analysis of AES(800)-1/2.5 after 30 min, 4 h and 24 h in SCPS.

High impedance modulus |Z| at low-frequency (LF) ranges indicate the level of porosity and defects present in the OIH films [69]. Generally, for all OIH samples, as the immersion time increases the impedance modulus decreases. For OIH samples AES(800)-1/2, Figure 9 shows that this value decreases as the immersion time in SCPS increases. The highest decrease (about two magnitude orders when compared to samples immersed for 30 min) was found after 4 h of immersion. After 24 h, the decrease is lower than one order of magnitude when compared to samples after 4 h of immersion. For OHI samples AES(800)-1/2.5 (Figure 10) the bode plots are similar after 4 h and 24 h, but decreased about one order of magnitude when compared to samples immersed for 30 min. For AES(800)-1/5 samples, shown in Figure 11, there is a clear decrease after 4 h of immersion. However, after 24 h of immersion, the impedance shows a similar behaviour when compared to samples immersed for 30 min. This may be explained by the lower ratio between the organic and inorganic amounts of precursors.

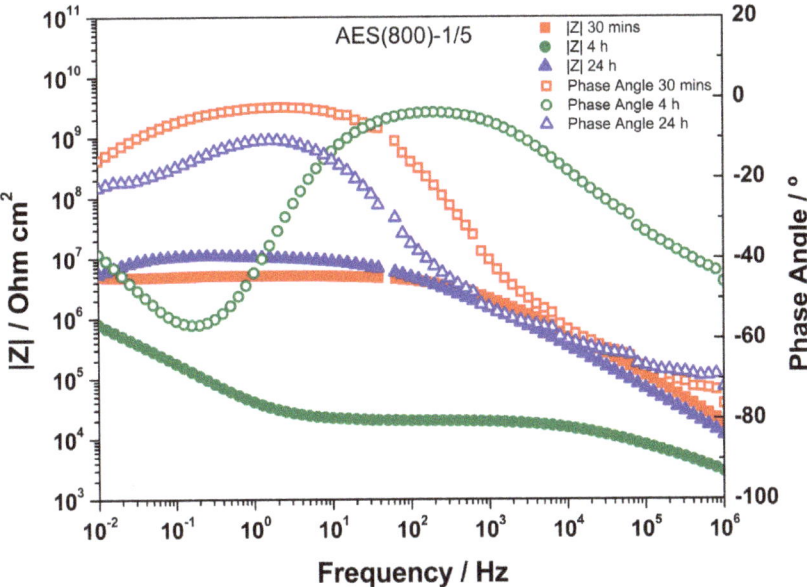

Figure 11. Bode plot for EIS analysis of AES(800)-1/5 after 30 min, 4 h and 24 h in SCPS.

Generally, the results show that the dielectric properties of the OIH materials change during the immersion in SCPS (pH > 12.5). Nevertheless, the |Z| values are within the range of 10^6–10^9. This behaviour suggests that the OIH materials synthesized show suitable properties to be embedded in fresh concrete. Furthermore, according to previous studies [68] materials with similar values were able to resist to the high pH of mortars.

3.4. TGA Analysis

Figure 12 shows that the highest degradation processes of the OIH materials occurred in the range between 400 and 600 °C, which is according to the literature [70–73]. Higher weight loss is shown for the sample with a smaller 3-APTES/PDMS ratio. These degradation processes are due to the depolymerization of the matrices and cleavage of Si–C bonds. The sample with the higher 3-APTES/PDMS ratio (AES(800)-1/5) has a smaller percentage of weight loss due to depolymerization. This may be explained by the decrease in the ratio between the organic and inorganic amounts of precursors. Table 3 shows the 5% weight loss temperature (T_5) and the temperature of the maximum rate of weight loss (T_{max}). It is shown that the higher the stoichiometric molar ratio between 3-ATPES and PDMS, the lower is the T_5 due to the higher volatile content. T_{max} (Table 4) does not show a clear association with this ratio. However, the maximum rate of weight loss is higher for smaller ratios between organic and inorganic amounts of precursors.

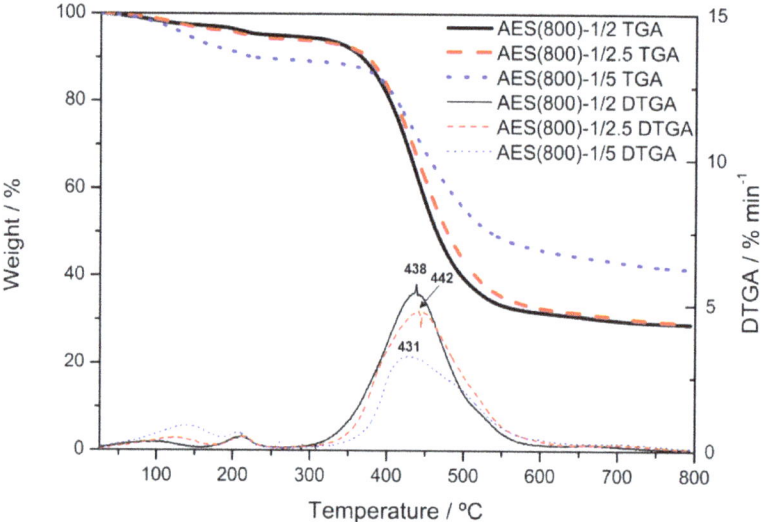

Figure 12. Experimental TGA and DTGA traces for AES(800)-1/2, AES(800)-1/2.5 and AES(800)-1/5 obtained at a heating rate of 10 °C/min.

Table 4. Data collected from TGA and DTGA traces.

Films	T_5	T_{max}
AES(800)-1/2	256	438
AES(800)-1/2.5	213	442
AES(800)-1/5	135	431

The TGA/DTGA data show that the main degradation processes occurred at temperatures above 350 °C; therefore, these OIH materials are suitable to apply in fresh concrete, since the maximum values achieved during the concrete curing process reach temperatures around 70 °C [74].

4. Conclusions

New OIH sol-gel materials based on PDMS(800)-GET and 3-APTES, with different ratios, were synthesized and were characterized by FTIR, UV–Vis spectroscopy, EIS and TGA. Phph-doped OIH films were also synthesized. FTIR spectroscopy showed that the synthesis of the precursor by the reaction between PDMS(800)-GET and 3-APTES was successful by the disappearance of the PDMS-characteristic epoxy group band, as were the hydrolysis and condensation of this precursor, as shown by the disappearance of alkoxy group's band. The presence of the indicator on the doped films was confirmed by UV–Vis spectroscopy.

Dielectric and thermal properties of the materials show that these materials resist high alkaline environments (SCPS, pH > 12.5). The impedance values allow one to conclude that for the OIH films, after immersion in SCPS, they do not decrease significantly. TGA data show that the produced OIH materials are stable enough to be used in fresh concrete, since the thermal degradation only occurs for values above the curing process of concrete (70 °C). The results indicate that these materials have potential to be used as sensing material and functionalize OFS to assess the pH of concrete. However, further studies must be conducted on functionalized OFS in SCPS in order to validate the potentiality of these OIH materials on OFS.

Author Contributions: Conceptualization, R.B.F., C.J.R.S. and R.P.C.L.S.; formal analysis, R.P.C.L.S., R.B.F, C.J.R.S. and S.P.G.C.; funding acquisition, R.B.F and C.J.R.S.; investigation, R.B.F., R.P.C.L.S. and B.F.; methodology, R.B.F and C.J.R.S.; writing—original draft preparation, R.P.C.L.S. and R.B.F.; writing—review and editing, R.B.F., M.A., S.P.G.C. and C.J.R.S.; project administration, R.B.F. and C.J.R.S; resources, C.J.R.S., R.B.F. and M.A.; supervision, R.B.F. and C.J.R.S. All authors have read and agreed to the published version of the manuscript.

Funding: This research was funded by the Program Budget COMPETE—Operational Program Competitiveness and Internationalization—COMPETE 2020, and the Lisbon Regional Operational Program (its FEDER component), and by the budget of FCT Foundation for Science and Technology, I.P, grant number POCI-01-0145-FEDER-031220.

Acknowledgments: The authors acknowledge the support of Centro de Química, CQUM, which is financed by national funds through the FCT Foundation for Science and Technology, I.P. under the project UID/QUI/00686/2016.

Conflicts of Interest: The authors declare no conflict of interest. The funders had no role in the design of the study; in the collection, analyses, or interpretation of data; in the writing of the manuscript, or in the decision to publish the results.

References

1. Schmidt, H. New Type of Non-Crystalline Solids between Inorganic and Organic Materials. *J. Non Cryst. Solids* **1985**, *73*, 681–691. [CrossRef]
2. Mackenzie, J.D. Structures and Properties of Ormosils. *J. Sol Gel Sci. Technol.* **1994**, *2*, 81–86. [CrossRef]
3. Figueira, R. Multifunctional and Smart Organic-Inorganic Hybrid Sol-Gel Coatings for Corrosion Protection Application. In *Advances in Smart Coatings And Thin Films for Future Industrial and Biomedical Engineering Applications*; Elsevier: Amsterdam, The Netherlands, 2019.
4. Figueira, R.B.; Fontinha, I.R.; Silva, C.J.R.; Pereira, E.V. Hybrid Sol-Gel Coatings: Smart and Green Materials for Corrosion Mitigation. *Coatings* **2016**, *6*, 12. [CrossRef]
5. Figueira, R.B.; Silva, C.J.R. Application of Sol–Gel Method to Synthesize Organic–Inorganic Hybrid Coatings to Minimize Corrosion in Metallic Substrates. In *Hybrid Organic-Inorganic Interfaces*; John Wiley & Sons, Ltd.: New York, NY, USA, 2017; pp. 355–412. [CrossRef]
6. Figueira, R.B.; Silva, C.J.R.; Pereira, E.V. Organic–Inorganic Hybrid Sol–Gel Coatings for Metal Corrosion Protection: A Review of Recent Progress. *J. Coat. Technol. Res.* **2015**, *12*, 1–35. [CrossRef]
7. Figueira, R.B.; Silva, C.J.R.; Pereira, E.V. Hybrid Sol–Gel Coatings for Corrosion Protection of Galvanized Steel in Simulated Concrete Pore Solution. *J. Coat. Technol. Res.* **2016**, *13*, 355–373. [CrossRef]
8. Boaretto, N.; Bittner, A.; Brinkmann, C.; Olsowski, B.-E.; Schulz, J.; Seyfried, M.; Vezzù, K.; Popall, M.; Di Noto, V. Highly Conducting 3D-Hybrid Polymer Electrolytes for Lithium Batteries Based on Siloxane Networks and Cross-Linked Organic Polar Interphases. *Chem. Mater.* **2014**, *26*, 6339–6350. [CrossRef]
9. Mujahid, A.; Lieberzeit, P.A.; Dickert, F.L. Chemical Sensors Based on Molecularly Imprinted Sol-Gel Materials. *Materials* **2010**, *3*, 2196–2217. [CrossRef]
10. Wright, J.; Sommerdijk, N.A.J.M. *Sol-Gel Materials: Chemistry and Applications*, 1st ed.; Gordon and Breach: London, UK, 2001.
11. Ianculescu, A.C.; Vasilescu, C.A.; Crisan, M.; Raileanu, M.; Vasile, B.S.; Calugaru, M.; Crisan, D.; Dragan, N.; Curecheriu, L.; Mitoseriu, L. Formation Mechanism and Characteristics of Lanthanum-Doped BaTiO3 Powders and Ceramics Prepared by the Sol–Gel Process. *Mater. Charact.* **2015**, *106*, 195–207. [CrossRef]
12. Kumar, P.; Yadav, A.K.; Joshi, A.G.; Bhattacharyya, D.; Jha, S.N.; Pandey, P.C. Influence of Li Co-Doping on Structural Property of Sol-Gel Derived Terbium Doped Zinc Oxide Nanoparticles. *Mater. Charact.* **2018**, *142*, 593–601. [CrossRef]
13. Brinker, C.J.; Scherer, G.W. *Sol-Gel Science: The Physics and Chemistry of Sol-Gel Processing*; Elsevier: Amsterdam, The Netherlands, 1990.
14. Hench, L.L.; West, J.K. The Sol-Gel Process. *Chem. Rev.* **1990**, *90*, 33–72. [CrossRef]
15. Livage, J. Basic Principles of Sol-Gel Chemistry. In *Sol-Gel Technologies for Glass Producers and Users*; Aegerter, M., Mennig, M., Eds.; Springer: New York, NY, USA, 2004.
16. Sakka, S.; Kozuka, H. Sol-Gel Processing. In *Handbook of Sol-Gel Science and Technology*; Springer Science & Business Media: Berlin, Germany, 2005.
17. Su, H.-Y.; Chen, P.-L.; Lin, C.-S. Sol–Gel Coatings Doped with Organosilane and Cerium to Improve the Properties of Hot-Dip Galvanized Steel. *Corros. Sci.* **2016**, *102*, 63–71. [CrossRef]

18. Tatsumisago, M.; Honjo, H.; Sakai, Y.; Minami, T. Proton-Conducting Silica-Gel Films Doped with a Variety of Electrolytes. *Solid State Ion.* **1994**, *74*, 105–108. [CrossRef]
19. Owens, G.J.; Singh, R.K.; Foroutan, F.; Alqaysi, M.; Han, C.-M.; Mahapatra, C.; Kim, H.-W.; Knowles, J.C. Sol-Gel Based Materials for Biomedical Applications. *Prog. Mater. Sci.* **2016**, *77*, 1–79. [CrossRef]
20. Nedeljko, P.; Turel, M.; Lobnik, A. Hybrid Sol-Gel Based Sensor Layers for Optical Determination of Biogenic Amines. *Sens. Actuators B Chem.* **2017**, *246*, 1066–1073. [CrossRef]
21. Choodum, A.; Kanatharana, P.; Wongniramaikul, W.; NicDaeid, N. A Sol–Gel Colorimetric Sensor for Methamphetamine Detection. *Sens. Actuators B Chem.* **2015**, *215*, 553–560. [CrossRef]
22. Hashem, S.G.; Elsaady, M.M.; Afify, H.G.; Omer, W.E.; Youssef, A.O.; El-Kemary, M.; Attia, M.S. Determination of Uric Acid in Serum Using an Optical Sensor Based on Binuclear Pd(II) 2-Pyrazinecarboxamide-Bipyridine Doped in a Sol Gel Matrix. *Talanta* **2019**, *199*, 89–96. [CrossRef]
23. Wang, B.; Anslyn, E.V. *Chemosensors: Principles, Strategies, and Applications*; John Wiley & Sons, Ltd.: New York, NY, USA, 2011. [CrossRef]
24. Jerónimo, P.C.A.; Araújo, A.N.; Conceição, B.S.M.; Montenegro, M. Optical Sensors and Biosensors Based on Sol–Gel Films. *Talanta* **2007**, *72*, 13–27. [CrossRef]
25. Nivens, D.A.; Zhang, Y.; Angel, S.M. A Fiber-Optic PH Sensor Prepared Using a Base-Catalyzed Organo-Silica Sol–Gel. *Anal. Chim. Acta* **1998**, *376*, 235–245. [CrossRef]
26. Nivens, D.A.; Schiza, M.V.; Angel, S.M. Multilayer Sol–Gel Membranes for Optical Sensing Applications: Single Layer PH and Dual Layer CO2 and NH3 Sensors. *Talanta* **2002**, *58*, 543–550. [CrossRef]
27. Riehle, N.; Thude, S.; Götz, T.; Kandelbauer, A.; Thanos, S.; Tovar, G.E.M.; Lorenz, G. Influence of PDMS Molecular Weight on Transparency and Mechanical Properties of Soft Polysiloxane-Urea-Elastomers for Intraocular Lens Application. *Eur. Polym. J.* **2018**, *101*, 190–201. [CrossRef]
28. Zhao, W.; Yang, J.; Guo, H.; Xu, T.; Li, Q.; Wen, C.; Sui, X.; Lin, C.; Zhang, J.; Zhang, L. Slime-Resistant Marine Anti-Biofouling Coating with PVP-Based Copolymer in PDMS Matrix. *Chem. Eng. Sci.* **2019**, *207*, 790–798. [CrossRef]
29. Romo-Uribe, A.; Santiago-Santiago, K.; Reyes-Mayer, A.; Aguilar-Franco, M. Functional PDMS Enhanced Strain at Fracture and Toughness of DGEBA Epoxy Resin. *Eur. Polym. J.* **2017**, *89*, 101–118. [CrossRef]
30. Eduok, U.; Faye, O.; Szpunar, J. Recent Developments and Applications of Protective Silicone Coatings: A Review of PDMS Functional Materials. *Prog. Org. Coat.* **2017**, *111*, 124–163. [CrossRef]
31. Zhu, D.; Handschuh-Wang, S.; Zhou, X. Recent Progress in Fabrication and Application of Polydimethylsiloxane Sponges. *J. Mater. Chem. A* **2017**, *5*, 16467–16497. [CrossRef]
32. Fujii, T. PDMS-Based Microfluidic Devices for Biomedical Applications. *Microelectron. Eng.* **2002**, *61–62*, 907–914. [CrossRef]
33. Tamayo, A.; Téllez, L.; Rubio, J.; Rubio, F.; Oteo, J.L. Effect of Reaction Conditions on Surface Properties of TEOS–TBOT–PDMS Hybrid Materials. *J. Sol Gel Sci. Technol.* **2010**, *55*, 94–104. [CrossRef]
34. Xu, F.; Li, D. Effect of the Addition of Hydroxyl-Terminated Polydimethylsiloxane to TEOS-Based Stone Protective Materials. *J. Sol Gel Sci. Technol.* **2013**, *65*, 212–219. [CrossRef]
35. Dirè, S. Sol-Gel Derived Polysiloxane-Oxide Hybrid Materials: Extent of Phase Interaction. *J. Sol Gel Sci. Technol.* **2003**, *26*, 285–290. [CrossRef]
36. Wolf, M.P.; Salieb-Beugelaar, G.B.; Hunziker, P. PDMS with Designer Functionalities—Properties, Modifications Strategies, and Applications. *Prog. Polym. Sci.* **2018**, *83*, 97–134. [CrossRef]
37. Gao, H.; Hu, H.; Zhao, Y.; Li, J.; Lei, M.; Zhang, Y. Highly-Sensitive Optical Fiber Temperature Sensors Based on PDMS/Silica Hybrid Fiber Structures. *Sens. Actuators Phys.* **2018**, *284*, 22–27. [CrossRef]
38. Wang, D.; Sheng, B.; Peng, L.; Huang, Y.; Ni, Z. Flexible and Optical Fiber Sensors Composited by Graphene and PDMS for Motion Detection. *Polymers* **2019**, *11*. [CrossRef] [PubMed]
39. Taheri, S. A Review on Five Key Sensors for Monitoring of Concrete Structures. *Constr. Build. Mater.* **2019**, *204*, 492–509. [CrossRef]
40. Figueira, R.B.; Sousa, R.; Coelho, L.; Azenha, M.; de Almeida, J.M.; Jorge, P.A.S.; Silva, C.J.R. Alkali-Silica Reaction in Concrete: Mechanisms, Mitigation and Test Methods. *Constr. Build. Mater.* **2019**, *222*, 903–931. [CrossRef]
41. Winter, N.B. *Understanding Cement: An Introduction to Cement Production, Cement Hydration and Deleterious Processes in Concrete*; Microanalysis Consultants: St. Ives, UK, 2012.

42. Parrott, L.J. *A Review of Carbonation in Reinforced Concrete: A Review Carried out by C & CA under a BRE Contract*; Cement and Concrete Association: Wexham Springs: Liverpool, UK, 1987.
43. Behnood, A.; Van Tittelboom, K.; De Belie, N. Methods for Measuring PH in Concrete: A Review. *Constr. Build. Mater.* **2016**, *105*, 176–188. [CrossRef]
44. Figueira, R.B. Electrochemical Sensors for Monitoring the Corrosion Conditions of Reinforced Concrete Structures: A Review. *Appl. Sci.* **2017**, *7*, 1157. [CrossRef]
45. Plusquellec, G.; Geiker, M.R.; Lindgård, J.; Duchesne, J.; Fournier, B.; De Weerdt, K. Determination of the PH and the Free Alkali Metal Content in the Pore Solution of Concrete: Review and Experimental Comparison. *Cem. Concr. Res.* **2017**, *96*, 13–26. [CrossRef]
46. Ferreira, M.; Castro-Camus, E.; Ottaway, D.; López-Higuera, J.; Feng, X.; Jin, W.; Jeong, Y.; Picqué, N.; Tong, L.; Reinhard, B.; et al. Roadmap on Optical Sensors. *J. Opt.* **2017**, *19*, 083001. [CrossRef]
47. Leung, C.K.Y. Fiber Optic Sensors in Concrete: The Future? *NDT E Int.* **2001**, *34*, 85–94. [CrossRef]
48. Dhouib, M.; Conciatori, D.; Sorelli, L. Optical Fiber Chloride Sensor for Health Monitoring of Structures in Cold Regions. *Cold Reg. Eng.* **2019**, 391–397. [CrossRef]
49. Basheer, P.A.M.; Grattan, K.; Sun, T.; Long, A.; Mcpolin, D.O.; Xie, W. Fiber Optic Chemical Sensor Systems for Monitoring PH Changes in Concrete. *Proc. SPIE Int. Soc. Opt. Eng.* **2004**, *5586*. [CrossRef]
50. Mcpolin, D.O.; Basheer, P.A.M.; Grattan, K.; Long, A.; Sun, T.; Xie, W. Preliminary Development and Evaluation of Fibre Optic Chemical Sensors. *J. Mater. Civ. Eng.* **2011**, *23*, 1200–1210. [CrossRef]
51. Sugama, T.; Brothers, L.E.; Weber, L. Acid-Resistant Polydimethylsiloxane Additive for Geothermal Well Cement in 150°C H2SO4 Solution. *Adv. Cem. Res.* **2003**, *15*, 35–44. [CrossRef]
52. Sidek, O.; Afzal, M.H.B. A Review Paper on Fiber-Optic Sensors and Application of PDMS Materials for Enhanced Performance. In Proceedings of the 2011 IEEE Symposium on Business, Engineering and Industrial Applications (ISBEIA), Langkawi, Malaysia, 25–28 September 2011; pp. 458–463. [CrossRef]
53. Sidek, O.; Kabir, S.; Afzal, M.H.B. Fiber Optic-Based Sensing Approach for Corrosion Detection. In Proceedings of the PIERS Proceedings, Suzhou, China, 12–16 September 2011.
54. Park, C.; Han, Y.; Joo, K.-I.; Lee, Y.W.; Kang, S.-W.; Kim, H.-R. Optical Detection of Volatile Organic Compounds Using Selective Tensile Effects of a Polymer-Coated Fiber Bragg Grating. *Opt. Express* **2010**, *18*, 24753–24761. [CrossRef] [PubMed]
55. Tan, C.H.; Shee, Y.G.; Yap, B.K.; Adikan, F.R.M. Fiber Bragg Grating Based Sensing System: Early Corrosion Detection for Structural Health Monitoring. *Sens. Actuators Phys.* **2016**, *246*, 123–128. [CrossRef]
56. Choi, J.-I.; Lee, Y.; Kim, Y.Y.; Lee, B.Y. Image-Processing Technique to Detect Carbonation Regions of Concrete Sprayed with a Phenolphthalein Solution. *Constr. Build. Mater.* **2017**, *154*, 451–461. [CrossRef]
57. Figueira, R.B.; Callone, E.; Silva, C.J.R.; Pereira, E.V.; Dirè, S. Hybrid Coatings Enriched with Tetraethoxysilane for Corrosion Mitigation of Hot-Dip Galvanized Steel in Chloride Contaminated Simulated Concrete Pore Solutions. *Materials* **2017**, *10*, 306. [CrossRef] [PubMed]
58. Sánchez, M.; Alonso, M.C.; Cecílio, P.; Montemor, M.F.; Andrade, C. Electrochemical and Analytical Assessment of Galvanized Steel Reinforcement Pre-Treated with Ce and La Salts under Alkaline Media. *Cem. Concr. Compos.* **2006**, *28*, 256–266. [CrossRef]
59. Recio, F.; Alonso, C.; Gaillet, L.; Sánchez Moreno, M. Hydrogen Embrittlement Risk of High Strength Galvanized Steel in Contact with Alkaline Media. *Corros. Sci.* **2011**, *53*, 2853–2860. [CrossRef]
60. Sakka, S.; Almeida, R.M. Characterization and Properties of Sol-Gel Materials and Products. In *Handbook of Sol-Gel Science and Technology*; Springer Science & Business Media: Berlin, Germany, 2005.
61. Arkles, B.; Larson, G. *Silicon Compounds: Silanes & Silicones*, 3rd ed.; Gelest Inc.: Morrisville, PA, USA, 2013.
62. Cai, D.; Neyer, A.; Kuckuk, R.; Heise, H.M. Raman, Mid-Infrared, near-Infrared and Ultraviolet–Visible Spectroscopy of PDMS Silicone Rubber for Characterization of Polymer Optical Waveguide Materials. *J. Mol. Struct.* **2010**, *976*, 274–281. [CrossRef]
63. Macdonald, J.R. *Impedance Spectroscopy Emphasizing Solid Materials and Analysis*; John Wiley and Sons, Ltd.: New York, NY, USA, 1987.
64. Orazem, M.E.; Tribollet, B. *Electrochemical Impedance Spectroscopy*; John Wiley and Sons, Ltd.: New York, NY, USA, 2008.
65. Barsoukov, E.; Macdonald, J.R. (Eds.) *Impedance Spectroscopy: Theory, Experiment, and Applications*, 3rd ed.; Wiley: Hoboken, NJ, USA, 2018.

66. Orazem, M.E.; Tribollet, B. The Electrochemical Society Series. In *Electrochemical Impedance Spectroscopy*; John Wiley & Sons, Ltd.: New York, NY, USA, 2017; pp. I–III. [CrossRef]
67. Jorcin, J.-B.; Orazem, M.E.; Pébère, N.; Tribollet, B. CPE Analysis by Local Electrochemical Impedance Spectroscopy. *Electrochimica Acta* **2006**, *51*, 1473–1479. [CrossRef]
68. Figueira, R.B.; Silva, C.J.R.; Pereira, E.V.; Salta, M.M. Alcohol-Aminosilicate Hybrid Coatings for Corrosion Protection of Galvanized Steel in Mortar. *J. Electrochem. Soc.* **2014**, *161*, C349–C362. [CrossRef]
69. Qian, M.; Mcintosh Soutar, A.; Tan, X.H.; Zeng, X.T.; Wijesinghe, S.L. Two-Part Epoxy-Siloxane Hybrid Corrosion Protection Coatings for Carbon Steel. *Thin Solid Films* **2009**, *517*, 5237–5242. [CrossRef]
70. Moreira, S.D.F.C.; Silva, C.J.R.; Prado, L.A.S.A.; Costa, M.F.M.; Boev, V.I.; Martín-Sánchez, J.; Gomes, M.J.M. Development of New High Transparent Hybrid Organic–Inorganic Monoliths with Surface Engraved Diffraction Pattern. *J. Polym. Sci. Part B Polym. Phys.* **2012**, *50*, 492–499. [CrossRef]
71. Thomas, T.H.; Kendrick, T.C. Thermal Analysis of Polydimethylsiloxanes. I. Thermal Degradation in Controlled Atmospheres. *J. Polym. Sci. Part A 2 Polym. Phys.* **1969**, *7*, 537–549. [CrossRef]
72. Ručigaj, A.; Krajnc, M.; Sebenik, U. Kinetic Study of Thermal Degradation of Polydimethylsiloxane: The Effect of Molecular Weight on Thermal Stability in Inert Atmosphere. *Polym. Sci.* **2017**, *3*. [CrossRef]
73. Gonzalez, J.; Iglio, R.; Barillaro, G.; Duce, C.; Tiné, M. Structural and Thermoanalytical Characterization of 3D Porous PDMS Foam Materials: The Effect of Impurities Derived from a Sugar Templating Process. *Polymers* **2018**, *10*, 616. [CrossRef]
74. Taylor, H.F.W.; Famy, C.; Scrivener, K.L. Delayed Ettringite Formation. *Cem. Concr. Res.* **2001**, *31*, 683–693. [CrossRef]

© 2020 by the authors. Licensee MDPI, Basel, Switzerland. This article is an open access article distributed under the terms and conditions of the Creative Commons Attribution (CC BY) license (http://creativecommons.org/licenses/by/4.0/).

Article
Organic and Inorganic PCL-Based Electrospun Fibers

Adrián Leonés [1,2], Alicia Mujica-Garcia [1,3], Marina Patricia Arrieta [1,4], Valentina Salaris [1], Daniel Lopez [1,2], José Maria Kenny [1,4] and Laura Peponi [1,2,*]

1. Instituto de Ciencia y Tecnología de Polímeros (ICTP-CSIC), C/Juan de la Cierva 3, 28006 Madrid, Spain; aleones@ictp.csic.es (A.L.); alicia.mujica@unipg.it (A.M.-G.); marrie06@ucm.es (M.P.A.); valen.salaris@gmail.com (V.S.); daniel.l.g@csic.es (D.L.); jose.kenny@unipg.it (J.M.K.)
2. Interdisciplinary Platform for Sustainable Plastics towards a Circular Economy—The Spanish National Research Council (SusPlast-CSIC), 28006 Madrid, Spain
3. Facultad de Óptica y Optometría, Universidad Complutense de Madrid (UCM), Arcos de Jalón 118, 28037 Madrid, Spain
4. Civil and Environmental Engineering Department, University of Perugia, Via G, Duranti 93, 06125 Perugia, Italy
* Correspondence: lpeponi@ictp.csic.es

Received: 30 March 2020; Accepted: 2 June 2020; Published: 10 June 2020

Abstract: In this work, different nanocomposite electrospun fiber mats were obtained based on poly(ε-caprolactone) (PCL) and reinforced with both organic and inorganic nanoparticles. In particular, on one side, cellulose nanocrystals (CNC) were synthesized and functionalized by "grafting from" reaction, using their superficial OH– group to graft PCL chains. On the other side, commercial chitosan, graphene as organic, while silver, hydroxyapatite, and fumed silica nanoparticles were used as inorganic reinforcements. All the nanoparticles were added at 1 wt% with respect to the PCL polymeric matrix in order to compare the different behavior of the woven no-woven nanocomposite electrospun fibers with a fixed amount of both organic and inorganic nanoparticles. From the thermal point of view, no difference was found between the effect of the addition of organic or inorganic nanoparticles, with no significant variation in the T_g (glass transition temperature), T_m (melting temperature), and the degree of crystallinity, leading in all cases to high crystallinity electrospun mats. From the mechanical point of view, the highest values of Young modulus were obtained when graphene, CNC, and silver nanoparticles were added to the PCL electrospun fibers. Moreover, all the nanoparticles used, both organic and inorganic, increased the flexibility of the electrospun mats, increasing their elongation at break.

Keywords: electrospinning; PCL; organic nanoparticles; inorganic nanoparticles

1. Introduction

Poly(ε-caprolactone) (PCL) is a biodegradable and biocompatible aliphatic polyester. It offers a unique combination of polyolefin-like mechanical properties and polyester-like hydrolyzability [1,2], providing good compatibility with a wide range of other polymers in both blends and copolymer forms in order to tune its properties [3–5]. PCL is also used as a polymeric matrix to obtain nanocomposites by adding micro and nanoparticles to improve its mechanical properties and its thermal stability [6], thus taking also the advantage of the strong ability of PCL-based materials to be processed in different forms, such as bulk, films, and fibers, among others [7]. In particular, PCL fibers can be obtained by following different methods, such as melt-spinning, dry-spinning, and electrospinning, depending on the desired fiber properties as well as on the final applications [8–10].

The electrospinning process is the most suitable technique for drug delivery and tissue engineering applications of PCL-based nanocomposite fibers [11–14]. The electrospinning process, belonging to the electro-hydrodynamic processing [14], can consistently produce a wide variety of woven non-woven mat starting from polymeric solutions exposed to high electric fields that offer large surface areas, small inter-fibrous pore size, and high porosity [15–18], with interest in the biomedical field since they mimic the structure of extracellular matrix [19]. Moreover, the electrospinning process is one of the most efficient, simple, and versatile processing techniques able to control micro and nanofibers' structural and functional properties through a relatively simple and cost-effective approach [20–24].

Furthermore, the development of nanocomposites is one of the most used approaches to improve PCL properties (i.e., thermal, mechanical, etc.) [6]. Among nanoparticles, cellulose derivatives have been considered optimal, reinforcing materials for biopolymers since they are bio-based, biodegradable, stiff, lightweight, and highly abundant in nature at low cost [22]. In particular, in the last years, cellulose nanocrystals (CNC), obtained from the acid hydrolysis of microcrystalline cellulose, have attracted the attention of many researchers [25,26]. However, CNC with high amounts of hydroxyl groups on their surface tends to aggregate, limiting improvements in the final properties of the nanocomposites [27–29]. Therefore, the homogeneous dispersion of high polarity CNC into the hydrophobic polymers matrices could be favored, modifying the nanocrystal surfaces by physical (i.e., use of surfactant) [30] or chemical modifications (*grafting*) [31]. The "grafting from" approach is particularly interesting since, in this method, different polymers can be directly grown in situ from the hydroxyl groups of the nanocellulose surface [31].

Chitosan is the second organic substance, most abundant in nature, after cellulose. Moreover, it has generated enormous interest due to its various advantages, such as easy availability, positive charge, renewability, biodegradability, biocompatibility, non-toxicity, and antimicrobial activity [32]. Another organic nanoparticle is graphene, which presents extraordinary mechanical properties, as high Young modulus and hardness and excellent flexibility are being considered as an effective reinforcement for high-performance nanocomposites [33,34].

On the other hand, PCL can be modified with inorganic nanoparticles, as hydroxyapatite, to improve its mechanical and biological properties for tissue engineering purposes. Hydroxyapatite is an inorganic component of bone with bioactivity and biocompatibility and shows good osteoconductivity and bone-bonding ability [35–37].

Finally, fumed silica is an inorganic and amorphous particle widely used at the industrial level mainly because it is a low-cost material obtained as a by-product of silicon [38]. Nanosized SiO_2, in general, improves the mechanical performance of the corresponding nanocomposite, even at low concentration as a consequence of their small particle size [38,39]. Besides, metal nanoparticles have been extensively used as reinforcing agents to produce nanocomposites with several functional properties. Silver is one of the most conductive metals in nature, and silver nanoparticles (Ag) are known as a powerful antimicrobial agent, with an outstanding broad-spectrum of antibacterial effects against both Gram-positive and Gram-negative bacteria, and this is why Ag have been widely used in biomedical applications [40,41].

Therefore, in general, woven no-woven electrospun fiber mats can find applications in tissue engineering since they mimic the microarchitecture of the extracellular matrix [19,20]. Their structure can improve cell adhesion, proliferation, migration, and differentiation. Moreover, PCL electrospun mats can be used as a drug release system. Similar works are reported with cellulose nanocrystal and graphene [42,43]. Among other applications, silver nanoparticles are used as reinforcement for PCL-based scaffold for wound dressing due to their high antibacterial properties [44]. At the same time, scaffolds with fibrous structure can be designed for nerve tissue regeneration due to their morphological similarity of the organization of neurons by developing electrospun chitosan-PCL scaffold [45]. The incorporation of HA is supposed to produce an ideal scaffold suitable for bone tissue because it improves osteoblast proliferation and differentiation [46]. Graphene can be added not only to improve the mechanical properties of PCL but also to obtain electrospun nanofibers with high antibacterial activity [47,48]. At the same time, silica nanoparticles can be used as an additive to enhance the mechanical properties and/or trigger an osteogenic response from osteoblastic progenitor cells for guided bone regeneration used for the treatment of lesions in the alveolar or mandible bone [49].

In this work, the processing conditions to obtain PCL-based electrospun fibers reinforced with organic as well as inorganic nanoparticles were optimized. Firstly, neat PCL and PCL reinforced with 1 wt% of silica nanoparticles were used to optimize the processing parameters. The optimized processing conditions were then used to produce PCL-based electrospun nanocomposite fibers in order to study the effect of the addition of different nanoparticles on their thermal and mechanical responses. In particular, the nanofillers used as reinforcements were CNC, CNC-g-PCL, chitosan, and graphene as organic nanoparticles, while HA, Ag, and SiO_2 were used as inorganic nanoparticles. All the nanoparticles were added to the PCL at 1 wt% in order to compare. The morphology, thermal behavior, and mechanical properties of the woven no-woven electrospun nanocomposite fibers were studied and compared with the neat PCL electrospun mat.

In particular, due to the intrinsic characteristic of both PCL polymer and the different nanoparticles, these electrospun nanocomposite mats could be considered for potential applications in bioactivity, such as antibacterial, biocompatibility, and biodegradability with proper mechanical performance.

2. Materials and Methods

2.1. Materials

Poly(ε-caprolactone) (PCL) (PCL CAPA 6500, Mn = 50,000 g/mol, 0.5 wt% ε-caprolactone monomer) was kindly donated by Perstorp (Malmö, Sweden).

Chitosan (degree of deacetylation > 75%) [17], microcrystalline cellulose, as well as commercial hydroxyapatite nanofillers (HA, the particle size of about 30 nm), were purchased from Sigma-Aldrich (Madrid, Spain). Silver nanoparticles, Ag, (P203, Cima NanoTech, Caesarea, Israel) were previously purified by a thermal treatment, obtaining a specific surface area of 4.9 m^2/g and particle size distribution from 20 to 70 nm [50]. Graphene nanoplatelets were supplied by Cheap Tubes Inc., Grafton, VT, USA (Grade 2) [51]. As indicated, they had a surface area of about 100 m^2/g and an average thickness of a bit less than 10 nm. They were used as supplied by the manufacturer without any functionalization process. Fumed silica dioxide nanopowder, SiO_2 (primary particle average size: 7–14 nm), was purchased from Interchim Innovations (Montluçon, France) [39]. Chloroform (99.6% purity) (CF) and dimethylformamide (DMF) (99.5% purity) were supplied by Sigma Aldrich (Madrid, Spain).

2.2. Synthesis of Cellulose Nanocrystals

Cellulose nanocrystals were obtained by sulfuric acid hydrolysis of 64% (wt/wt) of microcrystalline cellulose (MCC) stirring at 45 °C for 30 min [26,31]. The acid was further eliminated by centrifugation; the sediment was then dialyzed until neutral pH. An ion exchange resin was added to the cellulose suspension for 24 h, and it was then removed by filtration followed by ultrasonic treatment. Cellulose nanocrystal (CNC) solutions were then neutralized (1.0% (w/w) of 0.25 mol L^{-1} NaOH). Finally, the CNC solution was sonicated to get a stable suspension of the nanofillers.

2.3. CNC Functionalization with PCL Chains

CNC surface chemical modification was performed by grafting PCL chains onto the CNC surface by ring-opening polymerization (ROP) of ε-caprolactone (ε-CL) by using the surface hydroxyl groups of the CNC as initiator, as schematically shown in Scheme 1. The procedure for CNC functionalization was previously reported for PCL grafting [28,31]. Briefly, the aqueous suspension of CNC was solvent-exchanged with acetone, then with dichloromethane, and finally with previously dried toluene with phosphorus pentoxide. For each solvent exchange step, the solution was centrifuged and re-dispersed three times.

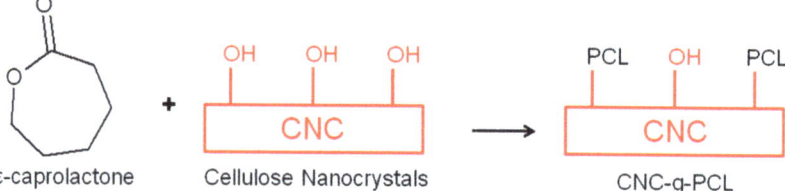

Scheme 1. CNC with grafted PCL chains. CNC, cellulose nanocrystal; PCL, poly(ε-caprolactone).

The CNC-g-PCL nanocrystals were observed by field emission scanning electron microscopy, FESEM, (Hitachi S8000). The structure of CNC-g-PCL was characterized by Fourier-transform infrared (FT–IR) in the attenuated total reflectance (ATR) mode (as well as Raman spectroscopy), and its thermal stability was studied by thermogravimetric analysis (TGA), as indicated in the characterization section.

2.4. Preparation of Electrospun Nanofibers

Electrospun fibers were prepared by means of a coaxial Electrospinner Yflow 2.2.D-XXX (Nanotechnology Solutions) with a vertical standard configuration equipped with two concentric needles and connected to a high voltage power. The polymer solution flew through the inner needle, and the solvent flew through the outer one.

PCL electrospun fibers, named ePCL, were prepared from polymer solutions of PCL (10 wt%) in a solvent mixture of $CHCl_3$:DMF (4:1) using a magnetic stirrer for 24 h at room temperature. The same solution was used to obtain PCL-based nanocomposite electrospun fibers by adding 1 wt% with respect to the polymer matrix of organic as well as inorganic nanoparticles, such as CNC, CNC-g-PCL, chitosan, graphene, Ag, HA, SiO_2.

To prepare electrospun nanocomposite fibers, nanoparticles were dispersed separately (1 wt%) in the same mix of solvents using a magnetic stirrer for 2 h and ultrasonication for 2 min. Finally, to achieve homogeneous dispersion of the nanoparticles in the dissolved matrix, the polymer solution and nanoparticles' suspensions were mixed, stirred (2 h), and sonicated (2 min) to form nanocomposite solutions with 1 wt% with respect to the matrix in the final composition.

The polymer solutions were flown through the inner needle, and the same mixture of solvent used for polymer solutions was flown through the outer needle. The applied positive and negative voltages were set at 10 and −10 kV, respectively. The polymer flow rate, as well as the solvent flow rate, were fixed at 1.0 mL/h. Electrospun mats were randomly collected in a grounded aluminum foil collector situated perpendicular at 15 cm from the charged spinneret. The obtained electrospun mats were vacuumed for 48 h in a vacuum chamber to eliminate residual solvents before testing.

2.5. Characterization Techniques

2.5.1. Vibrational Spectroscopy: Raman and FT-IR Spectra

Neat CNC and functionalized CNC were characterized by Raman spectra using a RenishawInVia Reflex Raman system. An optical microscope was coupled to the system. The Raman scattering was excited using a diode laser at a wavelength of 785 nm. The laser beam was focused on the sample with a 100 × 0.85 microscope objective. The laser power at the sample was 350 mW. The exposure was for 10 s, and there were 10 accumulations for the Raman measurements.

Fourier-transform infrared (FT–IR) spectra were obtained in the attenuated total reflectance (ATR) mode. The measurements were performed using a Spectrum One FTIR spectrometer Perkin Elmer equipped with an internal reflection element of diamond in the range of 650–4000 cm^{-1} with 1 cm^{-1} of resolution and an accumulation of 16 scans.

2.5.2. Scanning Electron Microscopy

The morphology of PCL-based electrospun nanocomposite fibers was analyzed by means of scanning electron microscopy (SEM). Thus, electrospun fibers were sputtered with a gold/palladium layer and observed by a PHILIPS XL30 scanning electron microscope (SEM). Fiber diameters were statistically calculated by means of Fib_thick software executable under image analysis platform Fiji based on ImageJ.

2.5.3. Thermal Analysis

TGA analysis was performed to study the thermal degradation of the electrospun samples. Thermogravimetric analysis (TGA) was performed in a TA-TGA Q500 thermal analyzer. Dynamic TGA experiments were performed under a nitrogen atmosphere (flow rate of 60 mL/min). Samples were heated from room temperature to 700 °C at 10 °C·min^{-1}. In this case, the maximum degradation temperature (T_{max}) was calculated from the first derivative of the TGA curves.

The thermal behavior, as well as the degree of crystallinity (X_c), were studied by differential scanning calorimetry (DSC). DSC experiments were conducted in a Mettler Toledo DSC822 instrument under a nitrogen atmosphere (50 mL/min). Sample weights of about 4 mg were sealed in aluminum pans and heated from −90 to 100 °C at a heating rate of 10 °C min^{-1}. The glass transition temperature (T_g) was taken at the midpoint of heat capacity changes. The melting temperature (T_m) and cold crystallization temperature (T_{cc}) were obtained from the first heating, and the degree of crystallinity (χc) was calculated through Equation (1):

$$\chi_c = 100\% \times \left[\frac{\Delta H_m - \Delta H_{cc}}{\Delta H_m^c} \right] \quad (1)$$

where ΔH_m is the melting enthalpy, ΔH_{cc} is the cold crystallization enthalpy, and ΔH_m^c is the melting heat associated with pure crystalline PCL (148 J g^{-1}) [5].

2.5.4. Mechanical Analysis

The mechanical properties of the PCL-based electrospun nanocomposite mats were investigated by tensile test measurements determined at room temperature in an Instron dynamometer (model 3366) equipped with a 100 N load cell, at a crosshead speed of 10 mm min^{-1} and initial length of 30 mm. Dog-bone samples were prepared from the mats, and, at least, five specimens were tested for each formulation. From these experiments, the Young Modulus—as the slope of the curve between 0% and 2% of deformation—the elongation at break, and the maximum strain reached were obtained.

3. Results

3.1. CNC Synthesis and Functionalization

First, the synthesis and the characterization of the different organic and inorganic nanoparticles were performed. In particular, neat cellulose nanocrystals (CNC) were synthesized and functionalized in our lab, while the other nanoparticles used in this work were purchased.

In particular, cellulose nanocrystals were synthesized in the laboratory by acid hydrolysis of commercial microcrystalline cellulose. Then, surface modification of the CNC was carried out by a "grafting from" reaction, grafting PCL chains onto the CNC surface by ring-opening polymerization (ROP) of –CL, using the surface hydroxyl groups of the CNC as initiator [28].

In particular, in order to ensure the presence of PCL chains in CNC-g-PCL during the "grafting from" reaction, vibrational spectroscopies studies were conducted, comparing the results obtained from neat PCL and CNC with CNC grafted with PCL (Figure 1). The FTIR spectrum of the CNC-g-PCL (Figure 1a) showed the characteristic peaks of CNC and of PCL. It was possible to observe a broad peak corresponding to the hydroxyl groups of CNC in the 3200 cm^{-1} to 3600 cm^{-1} region and a shaper peak of carboxyl groups of PCL at 1721 cm^{-1}, confirming the presence of both of them in the nanoparticles. Figure 1b shows the Raman spectrum of the PCL, CNC, and CNC-g-PCL, confirming the presence of PCL and CNC in the sample. These results confirmed the success of the grafting procedure.

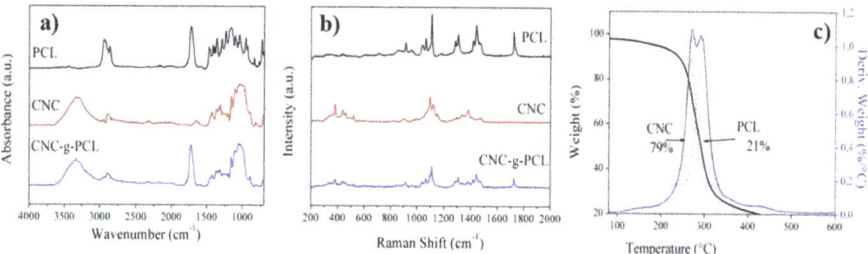

Figure 1. Vibrational spectroscopy: (**a**) FTIR spectra for PCL, CNC and CNC-g-PCL, (**b**) Raman spectra for PCL, CNC and CNC-g-PCL, and (**c**) thermogravimetric analysis of CNC-g-PCL. CNC-g-PCL: CNC with grafted PCL chains.

Moreover, TGA was used to determine the amount of CNC and PCL in the CNC-g-PCL nanoparticles. Figure 1c shows the weight loss and the derivative of weight loss for CNC-g-PCL. Two overlapped peaks were observed, in which one of them corresponded to the thermal degradation of CNC and the other one to the thermal degradation of PCL chains. CNC peak presented a maximum corresponding to 270 °C, and PCL showed the maximum peak at 292 °C. The amounts of the components were computed by fitting the curves with two Gaussian curves, so the amount of PCL grafted chains resulted as 21 wt%, while the amount of CNC was 79 wt%, in the range of already reported results [28].

3.2. Morphological Analysis

Morphological analysis was performed for all the organic and inorganic nanoparticles used and reported in Figure 2.

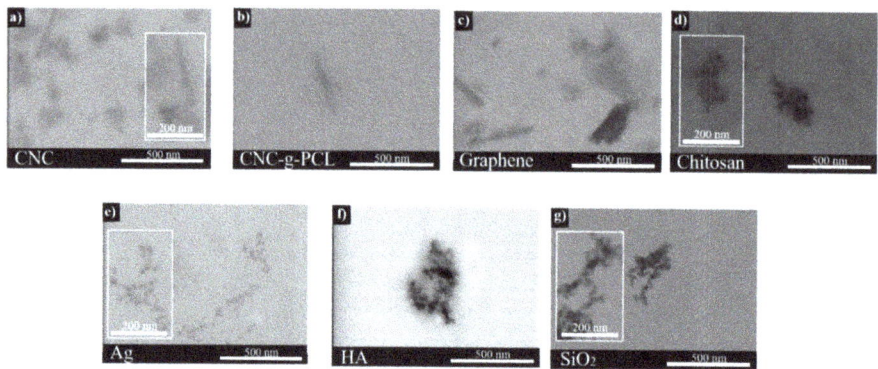

Figure 2. FE-SEM images of the different nanoparticles used in this work: (**a**) CNC, (**b**) CNC-g-PCL, (**c**) Graphene, (**d**) Chitosan, (**e**) silver (Ag), (**f**) hydroxyapatite (HA), (**g**) silica (SiO_2) nanoparticles.

From the FE-SEM images, the nanometer range size of the different nanoparticles was confirmed. In particular, the average length of CNC was determined by image analysis (ImageJ software), obtaining an average value of 179 ± 20 nm for the length and of 14.1 ± 1.6 nm for the diameter (Figure 2a), in good agreement with previous synthesized CNC [26]. Figure 2b shows individual nanoparticles of CNC-g-PCL, showing the characteristic rod-like morphology. The average length of CNC-g-PCL was determined by image analysis (ImageJ software), obtaining a value of 317 ± 59 nm for the length and a value of 24.4 ± 10.0 nm for the diameter, respectively, indicating an increase in both length and diameters due to the polymeric chain grafted onto the surface of the cellulose nanocrystals, in accordance with previous works [28].

Regarding the other nanoparticles used in the present work, the average dimension measured was in accordance with the values indicated by the seller, as indicated in Table 1. However, we could summarize that all the nanoparticles used in this work were smaller than 50 nanometers and could be easily used in the electrospinning process (Figure 2).

Table 1. Nanoparticles used and their dimensions.

Nanoparticles	Dimensions
CNC	179 ± 20 nm
	14.1 ± 1.6 nm
CNC-g-PCL	317 ± 59 nm
	24.4 ± 10.0 nm
Graphene	286.6 ± 64.6 nm
	35.5 ± 1.0 nm
Chitosan	23.7 ± 2.3 nm
Ag	22.4 ± 2.4 nm
HA	18.8 ± 2 nm
SiO_2	24.5 ± 2.9 nm

CNC: Cellulose nanocrystals; CNC-g-PCL: CNC with grafted PCL chains; HA: hydroxyapatite.

3.3. Optimization of the Electrospinning Process Parameters

Once the nanoparticles were characterized, we proceeded with the optimization of the electrospinning process and the characterization of the woven no-woven reinforced and no-reinforced PCL-based electrospun fiber mats. In particular, for the electrospinning process, PCL should be homogeneously dissolved in a proper solvent, and it is known that an effective solvent should present a similar solubility parameter (δ) to that of the polymer [52]. The solubility parameter of PCL is between 15.9 and 21.2 $MPa^{1/2}$ [53]. Chloroform (CF) has shown to be an effective solvent for PCL; meanwhile, dimethylformamide (DMF) is normally used to facilitate the electrospinning process. CF and DMF have solubility parameters of 19 $MPa^{1/2}$ and 24.9 $MPa^{1/2}$, respectively [15]. Accordingly, good solubility in CF and DMF for PCL should be expected.

The electrospinning processing conditions used to prepare electrospun fibers were selected, taking into account the optimization carried out for the production of neat PCL electrospun fibers (ePCL), as well as for silica-reinforced PCL electrospun fibers (ePCL + SiO_2). Both neat PCL electrospun fibers and PCL-based electrospun nanocomposite fibers were obtained from a PCL solution (10 wt%) and nanoparticle dispersions (1 wt%) in a mixture of solvents (chloroform-DMF 4:1). Firstly, the PCL solution was prepared, and, at the same time, nanofillers were dispersed separately. Then, the polymer solution and nanoparticles' suspensions were mixed, and then the final stable dispersion of nanoparticles into the PCL solution was obtained after sonication. Thus, the concentration of PCL solutions was set at 10 wt%, and the working distance between the needle and the collector was set at 15 cm [15,22]. The solvent and polymer flow rate, as well as the positive and negative voltage, were optimized according to the experimental results reported in Figure 3.

According to the fiber formations for both neat PCL and PCL-based electrospun nanocomposite fibers, reported in the table of Figure 3, it was worth noting that when small voltages were applied, the fiber formation was avoided. Both neat PCL and PCL-based electrospun fibers were formed upon increasing the voltage applied. However, for high flow rates of the polymer solution, typical defects, named beads, were obtained (see Figure 3). The best conditions for both neat and reinforced electrospun fibers were with a small flow rate in the solvent solution pump, a flow rate of 1 mL/h for the polymer solutions, and a high voltage (run 18). In particular, the electrospun fibers obtained with these conditions are reported in Figure 3 for both ePCL and ePCL reinforced with silica nanoparticles.

These results showed that to ensure the formation of the so-called Taylor cone, and hence the formation of electrospun fibers without defects, higher Coulomb forces were required to favor the elongation of the polymeric drops, and thus, higher voltage values were required. Therefore, we considered these conditions as optimum to obtain and to compare the properties of neat PCL as well as of PCL-based electrospun fibers reinforced with both organic and inorganic nanoparticles at 1 wt%.

Run	Q solvent (mL/h)	Q polymer (mL/h)	V+ (kV)	V- (kV)	Results
1	1	1	1	1	No fibers formation
2	1	1	4	4	No fibers formation
3	1	1	4	7	No fibers formation
4	1	1	4	10	No fibers formation
5	1	1	7	4	No fibers formation
6	1	1	7	7	No fibers formation
7	1	1	7	10	No fibers formation
8	1	1	10	7	Beads
9	5	5	10	10	Beads
10	0.5	5	10	10	Beads
11	1	5	10	10	Beads
12	3	3	10	10	Beads
13	1	3	10	10	Beads
14	5	1	10	10	Beads
15	3	1	10	10	Beads
16	1	1	10	10	Beads
17	1	1	10	4	Beads
18	**0.5**	**1**	**10**	**10**	**Fibers**
19	5	0.5	10	10	Beads
20	1	0.5	10	10	Beads
21	0.5	0.5	10	10	Beads

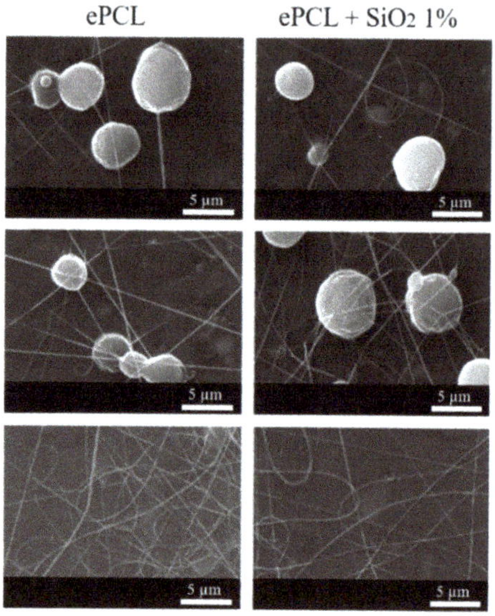

Figure 3. Optimization of the electrospinning process: processing-window parameters and SEM images for PCL and PCL reinforced with silica nanoparticles.

In accordance with the literature, we chose to study 1 wt% of different organic and inorganic nanoparticles, considering that this amount of nanoparticles could be effectively dispersed into the $CHCl_3$:DMF (4:1) solution, while higher concentrations produce a detriment of the structural and mechanical performance due to their reduced dispersion [22]. It was quite difficult to fix a unique amount of very disparate types of nanoparticles in order to compare the final properties of the nanocomposite electrospun fibers. Therefore, when we referred to the literature, in the case of CNC, we found that 1 wt% is a good amount to be dispersed into electrospun fibers [54]. Low amounts are also used for chitosan (i.e., 1 wt%) [17]. At the same time, Correa et al. developed electrospun scaffolds based on PCL reinforced with reduced graphite oxide (rGO) at concentrations up to 1 wt% [54]. Inorganic nanoparticles can be used also at low content. In fact, Ribeiro Nieto et al. studied PLA and PCL electrospun fibers reinforced with 1 wt% and 5 wt% of nano-sized hydroxyapatite, and the highest Young modulus was found for bionanocomposites reinforced with 1 wt% HA, also showing viable cells with early osteogenic activity [8]. Therefore, in this work, 1 wt% was the amount of both organic and inorganic nanoparticles added to the PCL electrospun mats.

Therefore, the electrospinning process was realized for all the systems, and the ePCL, as well as the PCL-based electrospun fiber mats, were obtained using the same processing window optimized before. Thus, the electrospun fibers based on PCL and reinforced with the organic, such as CNC and CNC-g-PCL, chitosan and graphene, as well as the other inorganic nanoparticles, such as silver and hydroxyapatite, were obtained. SEM images for all the PCL-based electrospun fiber mats are reported in Figure 4, and the corresponding average fiber diameters—for each system studied—are indicated on the top of each image.

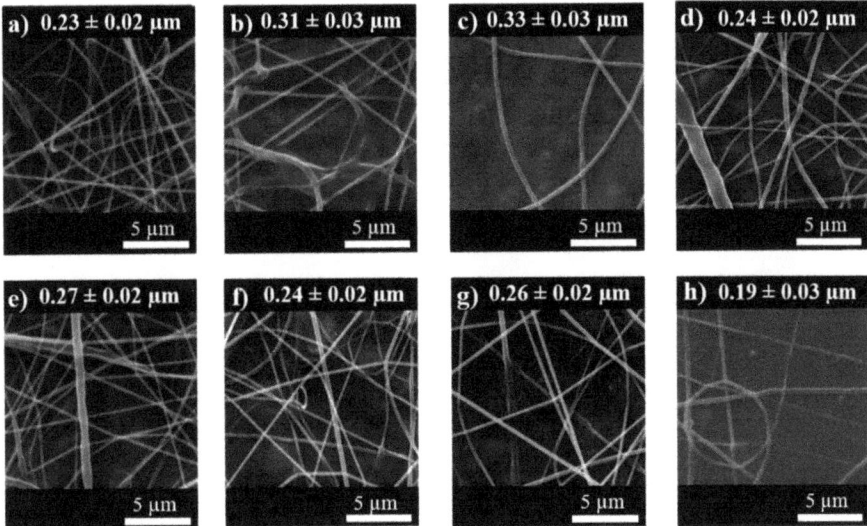

Figure 4. SEM images of (**a**) ePCL, (**b**) ePCL/CNC, (**c**) ePCL/CNC-g-PCL, (**d**) ePCL/Graphene, (**e**) ePCL/Chitosan, (**f**) ePCL/Ag, (**g**) ePCL/HA, (**h**) ePCL/SiO$_2$ electrospun fibers as well as their corresponding average diameters.

Thermal and mechanical characterization was also carried out. In particular, the main result of the thermal characterization is reported in Table 2, where the glass transition temperature (T_g), the melting temperature (T_m), and the degree of crystallinity (X_c) calculated from the DSC analysis and the maximum degradation temperature from TGA analysis are reported.

Table 2. Thermal characterization for the neat PCL and the reinforced PCL-based electrospun fibers.

Samples	T_g (°C)	T_m (°C)	X_c (%)	T_{max} (°C)
ePCL	−63	63	52	398
ePCL/CNC	−65	65	53	405
ePCL/CNC-g-PCL	−60	61	55	417
ePCL/Graphene	−62	62	50	406
ePCL/Chitosan	−62	64	50	417
ePCL/Ag	−62	64	48	400
ePCL/HA	−62	61	52	402
ePCL/SiO$_2$	−65	63	52	398

Electruspun PCL fibers (ePCL), ePCL reinforced with cellulose nanocrystals (ePCL/CNC), ePCL reinforced with CNC grafted with PCL chains (ePCL/CNC-g-PCL), ePCL reinforced with silver nanoparticles (ePCL/Ag), ePCL reinforced with hydroxyapatite (ePCL/HA), ePCL reinforced with fumed silica nanoparticles (ePCL/SiO2).

3.4. Thermal Analysis

In particular, from the thermal point of view, no significant differences in the T_g and T_m of the PCL-based electrospun fibers, as well as in their degree of crystallinity, were emerged from their comparison (Table 2). However, the degree of crystallinity of ePCL was quite high, 52%, indicating that the balance between the fiber formation and the solvent evaporation produced a quite crystalline material. Moreover, the addition of the nanoparticles did not alter the high degree of crystallinity of the neat ePCL. Different behavior, for example, was noted when working with other polymers, such as poly(lactic acid), where its neat electrospun fibers show a very low degree of crystallinity [23].

On the other hand, when we considered the maximum degradation temperature, it was evidenced that all the nanoparticles increased the maximum degradation temperature of the ePCL. The highest value was obtained when functionalized CNC and chitosan were added to the PCL. It seemed that the smallest increment was obtained when inorganic nanoparticles were added to the ePCL, considering that the addition of SiO_2 nanoparticles did not change the Tmax of neat ePCL, and the addition of both Ag and HA nanoparticles increased the Tmax of neat ePCL by about 1%.

For these reasons, we concluded that it was not possible to classify the different effects produced from the addition of organic and inorganic nanoparticles on the ePCL electrospun mats in terms of thermal characterization.

3.5. Mechanical Characterization

The mechanical characterization of all the different electrospun nanocomposite systems based on PCL was also performed by tensile test measurements (Figure 5a). First of all, it was important to point out that the measurement of "electrospun mats" was quite different from PCL "bulk" materials, considering the presence of fiber entanglements and the many microvoids between the fiber webs, producing very high error measurements, as indicated by the large error bars in Figure 5. Moreover, it was worth noting that the mechanical response of ePCL was quite different from the PCL bulk material, thus considering that when PCL was obtained in the form of electrospun fibers, the values of about 45% for the deformation at break were obtained. However, the ePCL electrospun nanocomposite fibers showed their reinforcing effects with respect to the neat ePCL in terms of Young modulus, tensile strength, as well as elongation at break, as indicated in Figure 5b–d, respectively.

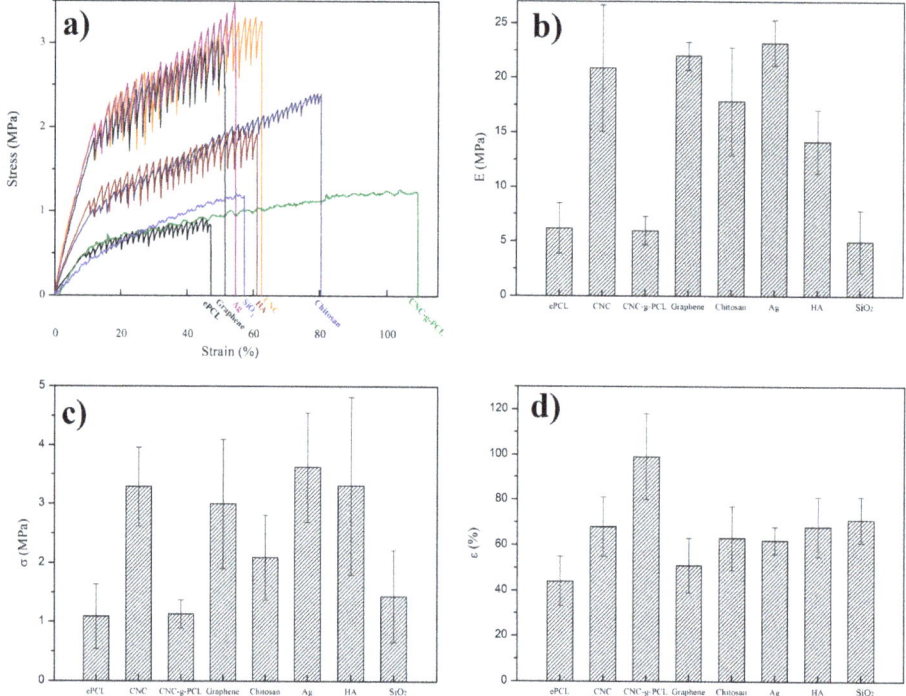

Figure 5. Mechanical response for the neat PCL and the reinforced PCL-based electrospun fibers. Stress-strain diagrams (**a**), Young modulus (**b**), tensile strength (**c**) and elongation at break (**d**) for all the samples studied.

Analyzing the different systems studied, both organic and inorganic nanofillers were able to increase the Young modulus and the tensile strength of electrospun PCL mat. The smallest effects in terms of Yong modulus and tensile strength were presented when CNC-g-PCL and SiO$_2$ were added to the PCL matrix; however, their elongation at break was strongly increased. In fact, the electrospun nanocomposite mat reinforced with CNC-g-PCL showed the highest elongation at break, with an increment of about 220% with respect to ePCL, suggesting that the better compatibilization between PCL and CNC was reached by grafting PCL chains onto the CNC surfaces. Regarding ePCL/SiO$_2$, although it was not able to increase the mechanical resistance of ePCL, it was able to increase the flexibility of the material by increasing the elongation at break by about 155% with respect to ePCL.

However, all the electrospun nanocomposite fibers showed improved elongation at break, with increments higher than 150% with respect to ePCL.

Furthermore, also in the case of the mechanical response, it was quite impossible to classify the reinforcement effect depending on the use of organic or inorganic nanoparticles. In fact, the addition of Ag nanoparticles, graphene, and CNC to the ePCL electrospun fibers provided an increment of more than 350% in terms of Young modulus with respect to ePCL, while the electrospun PCL fibers reinforced with chitosan and with HA showed an increment of about 250% with respect to the Young modulus of ePCL.

For the tensile strength, as said before, the addition of both CNC-g-PCL and SiO$_2$ nanoparticles slightly increased the tensile strength with respect to ePCL; however, the addition of CNC, graphene, Ag, and HA increased the tensile strength by about 300%.

Finally, in Figure 6, the variation of the properties of the woven no-woven electrospun nanocomposite systems with respect to the ePCL values has been summarized in order to better visualize the effect of the addition of the different organic and inorganic nanoparticles on the fiber diameter, degree of crystallinity, and the mechanical response.

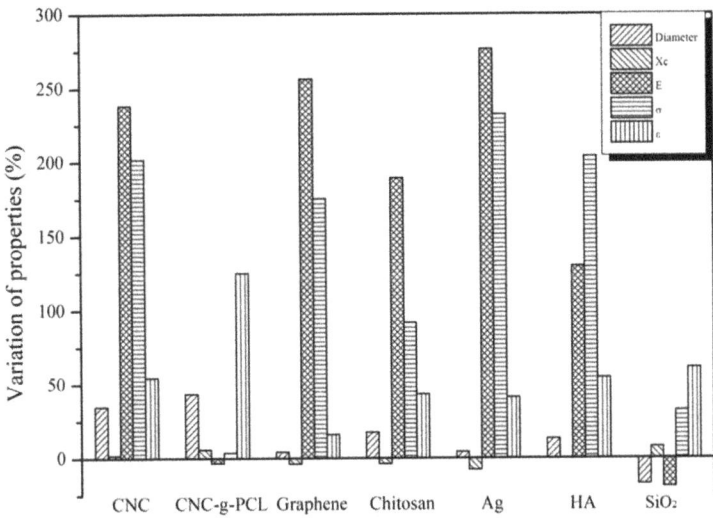

Figure 6. Variation of the properties of the woven no-woven electrospun nanocomposite systems with respect to the ePCL values.

4. Discussion

Summarizing, the values for the degree of crystallinity were quite similar, hindering a clear separation on the effect of the addition of organic or inorganic nanoparticles being strongly influenced by the high crystallinity of the neat ePCL electrospun fiber mats—higher than 50%.

Firstly, on comparing the properties of woven no-woven electrospun fibers and their corresponding bulk materials, we found that they were completely different. This was the case of PCL, but when we considered other polymers, such as PLA, the electrospun fiber mats presented a completely different mechanical response from bulk. For instance, PLA in bulk presents a Young modulus much higher and an elongation at break very smaller than PLA electrospun mats [55]. For PCL electrospun materials, the effect was different with respect to PLA-based electrospun mats. In fact, in our previous work [56], we reported the mechanical response of the same PCL obtained by the extrusion process, obtaining an elastic modulus of 294 ± 18 MPa, the maximum stress of 41 ± 4 MPa, and an elongation at break of 952 ± 23%. All the values obtained in bulk were much higher than the mechanical response obtained in electrospun fiber forms, where the elastic modulus obtained was about 6 MPa, the tensile strength was about 1 MPa, and the elongation at break was less than 50%. It was clear that PCL in the form of woven no-woven electrospun mats lost its capability of being a very flexible material.

It is expected that the elastic modulus and the tensile strength of the electrospun PCL-based fibers increase as the fiber diameter decreases [57]. Nevertheless, it could be observed that in the case of organic PCL-based electrospun fibers, the flexibility of the material increased with increasing the average fiber diameter. This behavior was particularly evident in the case of functionalized CNC (CNC-*g*-PCL), which showed a higher average fiber diameter and the higher elongation at break. Considering the inorganic nanofillers, it was observed that SiO_2 nanoparticles were able to produce the highest increment on the elongation at break among inorganic PCL-based electrospun fibers. However, it was quite difficult to differentiate the mechanical response of the PCL-based electrospun nanocomposite fibers regarding the effects of organic and inorganic nanofillers as a clear trend was not found.

It is well known that the different morphologies of the nanoparticles (0 D, 1 D, or 2 D dimensions) can strongly affect the mechanical response as well as the degree of crystallinity of the electrospun fibers [6]. However, in our case, it was difficult to differentiate the main results in terms of thermal or mechanical properties, depending on the geometry of the nanoparticles. In fact, the highest Young modulus and tensile strength were obtained with CNC, rod-like shaped Ag with spherical geometry, and graphene with a layered structure, indicating that the three different geometries provided the highest mechanical response.

5. Conclusions

Woven no-woven electrospun fibers based on PCL and reinforced with both organic and inorganic nanoparticles were obtained. As organic nanoparticles, we used cellulose nanocrystals, chitosan, and graphene, while, as inorganic nanoparticles, we used silver, hydroxyapatite, and fumed silica nanoparticles. In particular, cellulose nanocrystals (CNCs) were synthesized and functionalized by "*grafting from*" reaction, using their superficial OH– group to graft PCL chains. All the nanoparticles were added at 1 wt% with respect to the polymeric matrix, in order to be compared to each other. No difference was found between the effect of the addition of organic or inorganic nanoparticles on the thermal properties, considering that no significant variation in the T_g, T_m, and degree of crystallinity was obtained, leading to a high crystallinity electrospun mats with a degree of crystallinity of the electrospun PCL matrix of about 50%. However, all the nanoparticles increased the maximum degradation temperature of the respective nanocomposite electrospun fibers, indicating a good interaction between nanoparticles and the polymeric matrix. From the mechanical point of view, the highest values of Young modulus were obtained when graphene, CNC, and silver nanoparticles were added to the ePCL. On the other hand, all the nanoparticles used, both organic and inorganic types, increased the flexibility of the electrospun mats, increasing the elongation at break.

Author Contributions: Conceptualization, L.P.; Funding acquisition, L.P., J.M.K., and D.L.; Investigation, A.L., A.M.-G., and V.S.; Methodology, M.P.A., A.L., and A.M.-G.; Supervision, L.P. and J.M.K.; Writing—Original draft, L.P., A.M.-G., and M.P.A.; Writing—Review and editing, L.P., J.M.K., and D.L. All authors have read and agreed to the published version of the manuscript.

Funding: This work was funded by MAT2017-88123-P and PCIN-2017-036 (MINEICO FEDER EU).

Acknowledgments: Authors thank the Spanish Ministry of Economy, Industry, and Competitiveness (MINEICO) MAT2017-88123-P, POLYMAGIC: (PCIN-2017-036) co-financed with EU FEDER funds. M.A. and L.P. acknowledge the "Juan de la Cierva" (FJCI-2017-33536) and "Ramon y Cajal" (RYC-2014-15595) contracts from the MINEICO, respectively.

Conflicts of Interest: The authors declare no conflicts of interest.

References

1. Lee, K.H.; Kim, H.Y.; Khil, M.S.; Ra, Y.M.; Lee, D.R. Characterization of nano-structured poly(ε-caprolactone) nonwoven mats via electrospinning. *Polymer* **2003**, *44*, 1287–1294. [CrossRef]
2. Mohanty, A.K.; Misra, M.; Hinrichsen, G. Biofibres, biodegradable polymers and biocomposites: An overview. *Macromol. Mater. Eng.* **2000**, *276–277*, 1–24. [CrossRef]
3. Peponi, L.; Navarro-Baena, I.; Sonseca, A.; Gimenez, E.; Marcos-Fernandez, A.; Kenny, J.M. Synthesis and characterization of PCL-PLLA polyurethane with shape memory behavior. *Eur. Polym. J.* **2013**, *49*, 893–903. [CrossRef]
4. López-Rodríguez, N.; López-Arraiza, A.; Meaurio, E.; Sarasua, J.R. Crystallization, morphology, and mechanical behavior of polylactide/poly(ε-caprolactone) blends. *Polym. Eng. Sci.* **2006**, *46*, 1299–1308. [CrossRef]
5. Peponi, L.; Navarro-Baena, I.; Báez, J.E.; Kenny, J.M.; Marcos-Fernández, A. Effect of the molecular weight on the crystallinity of PCL-b-PLLA di-block copolymers. *Polymer* **2012**, *53*, 4561–4568. [CrossRef]
6. Peponi, L.; Puglia, D.; Torre, L.; Valentini, L.; Kenny, J.M. Processing of nanostructured polymers and advanced polymeric based nanocomposites. *Mater. Sci. Eng. R Rep.* **2014**, *85*, 1–46. [CrossRef]
7. Lamastra, F.R.; Puglia, D.; Monti, M.; Vella, A.; Peponi, L.; Kenny, J.M.; Nanni, F. Poly(ε-caprolactone) reinforced with fibres of Poly(methyl methacrylate) loaded with multiwall carbon nanotubes or graphene nanoplatelets. *Chem. Eng. J.* **2012**, *195–196*, 140–148. [CrossRef]
8. Ribeiro Neto, W.A.; Pereira, I.H.L.; Ayres, E.; De Paula, A.C.C.; Averous, L.; Góes, A.M.; Oréfice, R.L.; Suman Bretas, R.E. Influence of the microstructure and mechanical strength of nanofibers of biodegradable polymers with hydroxyapatite in stem cells growth. Electrospinning, characterization and cell viability. *Polym. Degrad. Stab.* **2012**, *97*, 2037–2051. [CrossRef]
9. Hutmacher, D.W.; Schantz, T.; Zein, I.; Ng, K.W.; Teoh, S.H.; Tan, K.C. Mechanical properties and cell cultural response of polycaprolactone scaffolds designed and fabricated via fused deposition modeling. *J. Biomed. Mater. Res.* **2001**, *55*, 203–216. [CrossRef]
10. Pal, J.; Kankariya, N.; Sanwaria, S.; Nandan, B.; Srivastava, R.K. Control on molecular weight reduction of poly(ε-caprolactone) during melt spinning—A way to produce high strength biodegradable fibers. *Mater. Sci. Eng. C* **2013**, *33*, 4213–4220. [CrossRef] [PubMed]
11. Keun Kwon, I.; Kidoaki, S.; Matsuda, T. Electrospun nano- to microfiber fabrics made of biodegradable copolyesters: Structural characteristics, mechanical properties and cell adhesion potential. *Biomaterials* **2005**, *26*, 3929–3939. [CrossRef] [PubMed]
12. Nematpour, N.; Farhadian, N.; Ebrahimi, K.S.; Arkan, E.; Seyedi, F.; Khaledian, S.; Shahlaei, M.; Moradi, S. Sustained release nanofibrous composite patch for transdermal antibiotic delivery. *Colloids Surfaces A Physicochem. Eng. Asp.* **2020**, *586*, 124267. [CrossRef]
13. Kai, D.; Liow, S.S.; Loh, X.J. Biodegradable polymers for electrospinning: Towards biomedical applications. *Mater. Sci. Eng. C* **2015**, *45*, 659–670. [CrossRef] [PubMed]
14. Cruz-Salas, C.N.; Prieto, C.; Calderón-Santoyo, M.; Lagarón, J.M.; Ragazzo-Sánchez, J.A. Micro-and nanostructures of agave fructans to stabilize compounds of high biological value via electrohydrodynamic processing. *Nanomaterials* **2019**, *9*, 1659. [CrossRef] [PubMed]
15. Mujica-Garcia, A.; Navarro-Baena, I.; Kenny, J.M.; Peponi, L. Influence of the Processing Parameters on the Electrospinning of Biopolymeric Fibers. *J. Renew. Mater.* **2014**, *2*, 23–34. [CrossRef]
16. Torres-Giner, S.; Wilkanowicz, S.; Melendez-Rodriguez, B.; Lagaron, J.M. Nanoencapsulation of Aloe vera in Synthetic and Naturally Occurring Polymers by Electrohydrodynamic Processing of Interest in Food Technology and Bioactive Packaging. *J. Agric. Food Chem.* **2017**, *65*, 4439–4448. [CrossRef] [PubMed]
17. Arrieta, M.P.; López, J.; López, D.; Kenny, J.M.; Peponi, L. Effect of chitosan and catechin addition on the structural, thermal, mechanical and disintegration properties of plasticized electrospun PLA-PHB biocomposites. *Polym. Degrad. Stab.* **2016**, *132*, 145–156. [CrossRef]

18. Torres-Martínez, E.J.; Pérez-González, G.L.; Serrano-Medina, A.; Grande, D.; Vera-Graziano, R.; Cornejo-Bravo, J.M.; Villarreal-Gómez, L.J. Drugs Loaded into Electrospun Polymeric Nanofibers for Delivery. *J. Pharm. Pharm. Sci.* **2019**, *22*, 313–331. [CrossRef] [PubMed]
19. De Cassan, D.; Becker, A.; Glasmacher, B.; Roger, Y.; Hoffmann, A.; Gengenbach, T.R.; Easton, C.D.; Hänsch, R.; Menzel, H. Blending chitosan-g-poly(caprolactone) with poly(caprolactone) by electrospinning to produce functional fiber mats for tissue engineering applications. *J. Appl. Polym. Sci.* **2020**, *137*, 1–11. [CrossRef]
20. Kriegel, C.; Arrechi, A.; Kit, K.; McClements, D.J.; Weiss, J. Fabrication, functionalization, and application of electrospun biopolymer nanofibers. *Crit. Rev. Food Sci. Nutr.* **2008**, *48*, 775–797. [CrossRef] [PubMed]
21. Huang, Z.M.; Zhang, Y.Z.; Kotaki, M.; Ramakrishna, S. A review on polymer nanofibers by electrospinning and their applications in nanocomposites. *Compos. Sci. Technol.* **2003**, *63*, 2223–2253. [CrossRef]
22. Arrieta, M.P.; López, J.; López, D.; Kenny, J.M.; Peponi, L. Biodegradable electrospun bionanocomposite fibers based on plasticized PLA–PHB blends reinforced with cellulose nanocrystals. *Ind. Crops Prod.* **2016**, *93*, 290–301. [CrossRef]
23. Leonés, A.; Sonsera, A.; López, D.; Fiori, S.; Peponi, L. Shape memory effect on electrospun PLA-based fibers tailoring their thermal response. *Eur. Polym. J.* **2019**, *117*, 217–226. [CrossRef]
24. Sessini, V.; López Galisteo, A.J.; Leonés, A.; Ureña, A.; Peponi, L. Sandwich-Type Composites Based on Smart Ionomeric Polymer and Electrospun Microfibers. *Front. Mater.* **2019**, *6*, 1–15. [CrossRef]
25. Habibi, Y.; Lucia, L.A.; Rojas, O.J. Cellulose nanocrystals: Chemistry, self-assembly, and applications. *Chem. Rev.* **2010**, *110*, 3479–3500. [CrossRef] [PubMed]
26. Navarro-Baena, I.; Kenny, J.M.; Peponi, L. Thermally-activated shape memory behaviour of bionanocomposites reinforced with cellulose nanocrystals. *Cellulose* **2014**, *21*, 4231–4246. [CrossRef]
27. Mujica-Garcia, A.; Hooshmand, S.; Skrifvars, M.; Kenny, J.M.; Oksman, K.; Peponi, L. Poly(lactic acid) melt-spun fibers reinforced with functionalized cellulose nanocrystals. *RSC Adv.* **2016**, *6*, 9221–9231. [CrossRef]
28. Sessini, V.; Navarro-Baena, I.; Arrieta, M.P.; Dominici, F.; López, D.; Torre, L.; Kenny, J.M.; Dubois, P.; Raquez, J.M.; Peponi, L. Effect of the addition of polyester-grafted-cellulose nanocrystals on the shape memory properties of biodegradable PLA/PCL nanocomposites. *Polym. Degrad. Stab.* **2018**, *152*, 126–138. [CrossRef]
29. Miao, C.; Hamad, W.Y. Cellulose reinforced polymer composites and nanocomposites: A critical review. *Cellulose* **2013**, *20*, 2221–2262. [CrossRef]
30. Bondeson, D.; Oksman, K. Dispersion and characteristics of surfactant modified cellulose whiskers nanocomposites. *Compos. Interfaces* **2007**, *14*, 617–630. [CrossRef]
31. Paquet, O.; Krouit, M.; Bras, J.; Thielemans, W.; Belgacem, M.N. Surface modification of cellulose by PCL grafts. *Acta Mater.* **2010**, *58*, 792–801. [CrossRef]
32. Fernández-Pan, I.; Maté, J.I.; Gardrat, C.; Coma, V. Effect of chitosan molecular weight on the antimicrobial activity and release rate of carvacrol-enriched films. *Food Hydrocoll.* **2015**, *51*, 60–68. [CrossRef]
33. Geim, A.K.; Novoselov, K.S. The rise of graphene. *Nat. Mater.* **2007**, *6*, 183–191. [CrossRef] [PubMed]
34. Peponi, L.; Tercjak, A.; Verdejo, R.; Lopez-Manchado, M.A.; Mondragon, I.; Kenny, J.M. Confinement of functionalized graphene sheets by triblock copolymers. *J. Phys. Chem. C* **2009**, *113*, 17973–17978. [CrossRef]
35. Spadaccio, C.; Rainer, A.; Trombetta, M.; Vadalá, G.; Chello, M.; Covino, E.; Denaro, V.; Toyoda, Y.; Genovese, J.A. Poly-l-lactic acid/hydroxyapatite electrospun nanocomposites induce chondrogenic differentiation of human MSC. *Ann. Biomed. Eng.* **2009**, *37*, 1376–1389. [CrossRef] [PubMed]
36. Peponi, L.; Sessini, V.; Arrieta, M.P.; Navarro-Baena, I.; Sonseca, A.; Dominici, F.; Gimenez, E.; Torre, L.; Tercjak, A.; López, D.; et al. Thermally-activated shape memory effect on biodegradable nanocomposites based on PLA/PCL blend reinforced with hydroxyapatite. *Polym. Degrad. Stab.* **2018**, *151*, 36–51. [CrossRef]
37. Sonseca, A.; Peponi, L.; Sahuquillo, O.; Kenny, J.M.; Giménez, E. Electrospinning of biodegradable polylactide/hydroxyapatite nanofibers: Study on the morphology, crystallinity structure and thermal stability. *Polym. Degrad. Stab.* **2012**, *97*, 2052–2059. [CrossRef]
38. Salgado, C.; Arrieta, M.P.; Peponi, L.; Fernández-García, M.; López, D. Silica-nanocomposites of photo-crosslinkable poly(urethane)s based on poly(ε-caprolactone) and coumarin. *Eur. Polym. J.* **2017**, *93*, 21–32. [CrossRef]
39. Sessini, V.; Brox, D.; López, A.J.; Ureña, A.; Peponi, L. Thermally activated shape memory behavior of copolymers based on ethylene reinforced with silica nanoparticles. *Nanocomposites* **2018**, *4*, 19–35. [CrossRef]

40. Xu, X.; Yang, Q.; Wang, Y.; Yu, H.; Chen, X.; Jing, X. Biodegradable electrospun poly(l-lactide) fibers containing antibacterial silver nanoparticles. *Eur. Polym. J.* **2006**, *42*, 2081–2087. [CrossRef]
41. Gangadharan, D.; Harshvardan, K.; Gnanasekar, G.; Dixit, D.; Popat, K.M.; Anand, P.S. Polymeric microspheres containing silver nanoparticles as a bactericidal agent for water disinfection. *Water Res.* **2010**, *44*, 5481–5487. [CrossRef] [PubMed]
42. Hivechi, A.; Bahrami, S.H.; Siegel, R.A. Drug release and biodegradability of electrospun cellulose nanocrystal reinforced polycaprolactone. *Mater. Sci. Eng. C* **2019**, *94*, 929–937. [CrossRef] [PubMed]
43. Scaffaro, R.; Maio, A.; Botta, L.; Gulino, E.F.; Gulli, D. Tunable release of Chlorhexidine from Polycaprolactone-based filaments containing graphene nanoplatelets. *Eur. Polym. J.* **2019**, *110*, 221–232. [CrossRef]
44. Zhang, M.; Lin, H.; Wang, Y.; Yang, G.; Zhao, H.; Sun, D. Fabrication and durable antibacterial properties of 3 D porous wet electrospun RCSC/PCL nanofibrous scaffold with silver nanoparticles. *Appl. Surf. Sci.* **2017**, *414*, 52–62. [CrossRef]
45. Cooper, A.; Bhattarai, N.; Zhang, M. Fabrication and cellular compatibility of aligned chitosan-PCL fibers for nerve tissue regeneration. *Carbohydr. Polym.* **2011**, *85*, 149–156. [CrossRef]
46. Wutticharoenmongkol, P.; Sanchavanakit, N.; Pavasant, P.; Supaphol, P. Preparation and characterization of novel bone scaffolds based on electrospun polycaprolactone fibers filled with nanoparticles. *Macromol. Biosci.* **2006**, *6*, 70–77. [CrossRef] [PubMed]
47. Ramazani, S.; Karimi, M. Study the molecular structure of poly(ε-caprolactone)/graphene oxide and graphene nanocomposite nanofibers. *J. Mech. Behav. Biomed. Mater.* **2016**, *61*, 484–492. [CrossRef] [PubMed]
48. Liu, S.; Zeng, T.H.; Hofmann, M.; Burcombe, E.; Wei, J.; Jiang, R.; Kong, J.; Chen, Y. Antibacterial activity of graphite, graphite oxide, graphene oxide, and reduced graphene oxide: Membrane and oxidative stress. *ACS Nano* **2011**, *5*, 6971–6980. [CrossRef] [PubMed]
49. Castro, A.G.B.; Diba, M.; Kersten, M.; Jansen, J.A.; van den Beucken, J.J.J.P.; Yang, F. Development of a PCL-silica nanoparticles composite membrane for Guided Bone Regeneration. *Mater. Sci. Eng. C* **2018**, *85*, 154–161. [CrossRef] [PubMed]
50. Peponi, L.; Tercjak, A.; Torre, L.; Kenny, J.M.; Mondragon, I. Morphological analysis of self-assembled SIS block copolymer matrices containing silver nanoparticles. *Compos. Sci. Technol.* **2008**, *68*, 1631–1636. [CrossRef]
51. Monti, M.; Rallini, M.; Puglia, D.; Peponi, L.; Torre, L.; Kenny, J.M. Morphology and electrical properties of graphene-epoxy nanocomposites obtained by different solvent assisted processing methods. *Compos. Part A Appl. Sci. Manuf.* **2013**, *46*, 166–172. [CrossRef]
52. Arrieta, M.P.; López, J.; López, D.; Kenny, J.M.; Peponi, L. Development of flexible materials based on plasticized electrospun PLA-PHB blends: Structural, thermal, mechanical and disintegration properties. *Eur. Polym. J.* **2015**, *73*, 433–446. [CrossRef]
53. Bordes, C.; Fréville, V.; Ruffin, E.; Marote, P.; Gauvrit, J.Y.; Briançon, S.; Lantéri, P. Determination of poly(ε-caprolactone) solubility parameters: Application to solvent substitution in a microencapsulation process. *Int. J. Pharm.* **2010**, *383*, 236–243. [CrossRef] [PubMed]
54. Bellani, C.F.; Pollet, E.; Hebraud, A.; Pereira, F.V.; Schlatter, G.; Avérous, L.; Bretas, R.E.S.; Branciforti, M.C. Morphological, thermal, and mechanical properties of poly(ε-caprolactone)/poly(ε-caprolactone)-grafted-cellulose nanocrystals mats produced by electrospinning. *J. Appl. Polym. Sci.* **2016**, *133*, 4–11. [CrossRef]
55. Arrieta, M.P.; Perdiguero, M.; Fiori, S.; Kenny, J.M.; Peponi, L. Biodegradable electrospun PLA-PHB fibers plasticized with oligomeric lactic acid. *Polym. Degrad. Stab.* **2020**, 109226. [CrossRef]
56. Navarro-Baena, I.; Sessini, V.; Dominici, F.; Torre, L.; Kenny, J.M.; Peponi, L. Design of biodegradable blends based on PLA and PCL: From morphological, thermal and mechanical studies to shape memory behavior. *Polym. Degrad. Stab.* **2016**, *132*, 97–108. [CrossRef]
57. Wong, S.C.; Baji, A.; Leng, S. Effect of fiber diameter on tensile properties of electrospun poly(ε-caprolactone). *Polymer* **2008**, *49*, 4713–4722. [CrossRef]

© 2020 by the authors. Licensee MDPI, Basel, Switzerland. This article is an open access article distributed under the terms and conditions of the Creative Commons Attribution (CC BY) license (http://creativecommons.org/licenses/by/4.0/).

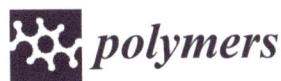

Article

Rheological, Mechanical and Morphological Characterization of Fillers in the Nautical Field: The Role of Dispersing Agents on Composite Materials

Silvia Vita [1,2], Rico Ricotti [2], Andrea Dodero [1,*], Silvia Vicini [1,*], Per Borchardt [3], Emiliano Pinori [3] and Maila Castellano [1]

1. Department of Chemistry and Industrial Chemistry, University of Genoa, Via Dodecaneso 31, 16146 Genova, Italy; silvia.vita@boero.it (S.V.); maila.castellano@unige.it (M.C.)
2. Boero Bartolomeo S.p.A., R&D "Riccardo Cavalleroni", Strada Comunale Savonesa 9, PST—Blocco F, Rivalta Scrivia, 15057 Tortona, Italy; r.ricotti@boero.it
3. Bioscience and Materials, RISE Research Institute of Sweden, Lindholmspiren 7 A, 41756 Göteborg, Sweden Borås, Sweden; per.borchardt@ri.se (P.B.); emiliano.pinori@ri.se (E.P.)
* Correspondence: andrea.dodero@edu.unige.it (A.D.); silvia.vicini@unige.it (S.V.); Tel.: +39-010-353-8726 (A.D.); +39-010-353-8713 (S.V.)

Received: 25 May 2020; Accepted: 9 June 2020; Published: 12 June 2020

Abstract: Coatings have a fundamental role in covering the external surface of yachts by acting both as protective and aesthetic layers. In particular, fillers represent the essential layer from the point of view of mechanical properties and consist of a polymeric matrix, different extenders and additives, and dispersing agents, with the latter having the role to provide good extender-matrix compatibility. In the present work, the effects of dispersing agents with an ionic or steric action on the interactions between hollow glass microspheres and an epoxy-polyamide resin are evaluated. Un-crosslinked filler materials are studied via rheological tests, whereas the mechanical and morphological properties of the crosslinked samples are assessed. The results clearly indicate that steric dispersing agents provide a much greater compatibility effect compared to ionic ones, owing to their steric hindrance capability, thus leading to better-performing filler materials with a less-marked Payne effect, which is here proved to be an efficient tool to provide information concerning the extent of component interactions in nautical fillers. To the best of our knowledge, this work represents the first attempt to deeply understand the role of dispersing agents, which are until now empirically used in the preparation of fillers.

Keywords: nautical fillers; extender-matrix interactions; dispersing agents; mechanical properties; rheological properties; Payne effect; morphological characterization

1. Introduction

Coating systems play a fundamental role in the construction of yachts and superyachts [1–4]. Indeed, by covering these external surfaces, coatings are most exposed to aggressive environments, such as seawater, marine atmosphere and thermal variations, thus providing a significant protection effect. Additionally, coatings must provide aesthetic properties typical of luxury products (e.g., brightness, light reflection, durability over time) [5–8]. Such performances are achieved through complex multilayer structures, namely painting systems. On the metallic substrate above the waterline, several layers with different thicknesses and functions are usually present. In particular, a bottom primer layer is covered by the filler and finishing filler strata, upon which another primer layer is applied before the undercoat and topcoat. The mechanical resistance of such a complex structure is mainly attributed to the filler/plaster layer, consisting in a composite material with a thickness of around 2 cm. Thus, it is not surprising that both the physicochemical features and the application of this layer are crucial for smoothing the surface, filling possible defects or voids and contributing to isolate the hulls [9]. However, few studies are available in the literature for these specific materials [5,6,10,11]. Fillers are usually made of two different parts consisting in an A component (e.g., epoxy resin) and a B component (e.g., curing agent based on a polyamide group), which, once mixed together in opportune ratios, form the final composite materials to be applied. In addition to the polymeric matrix, different types of additives, extenders and pigments are present in the formulation and require to be well dispersed in the matrix in order to exploit their functions [12,13]. The most common additives are rheological modifiers, antifoams and dispersing agents [5,12,14–17]. Rheological additives act on the viscosity of the samples, allowing to increase or reduce their tendency to flow and avoid the sagging phenomenon during the application step (i.e., thixotropic recovery), whereas anti-foams are used to obviate the formation of foams during the dispersion of extenders in the matrix phase [18–20]. The complexity of paint formulations is caused not only by their multicomponent composition but also by their multiphase nature and related thermodynamic instability [21–24]. In such a complex system, dispersing agents are essential to improve the incorporation of powders in the filler and ensure their stability during manufacturing, storage and application processes. The dispersing step is the most difficult and time/energy-consuming part of the entire paint manufacturing process, owing to the difference in surface tension between liquids (polymers and optional solvents) and powders (pigments and/or extenders) [25–27]. Dispersing agents are able to coat suspended powder particles to form a barrier that, either by ionic repulsion (i.e., an ionic dispersant generally having a low molecular weight) and/or steric hindrance (i.e., a non-ionic dispersant generally having a high molecular weight), prevents particle–particle interactions and aggregation. These agents, in comparison to the surfactants, are chemical compounds consisting in two well-defined parts: the oil soluble one (hydrophobic), with aliphatic or aromatic hydrocarbon residues, and the water soluble one (hydrophilic). The hydrophilic group can be ionic or non-ionic. In the first case, the stabilization mechanism is based on ionic repulsions with the formation of an electric layer (i.e., Helmholtz layer) on the particle surface, leading to electrostatic repulsive forces to guard against aggregation. Contrariwise, if the dispersing agents have a non-ionic nature, the stabilization mechanism is based on steric hindrance; these dispersants usually have pendant anchoring groups that are adsorbed onto the particle surface by hydrogen bonding, dipole–dipole interactions or Van der Waals forces. The free part of the chains is large enough to cause steric stabilization and to act as a bumper preventing the approach of particles to each other [28–33]. Concerning the extenders, whose function is mainly to reduce the density and cost of the fillers while maintaining satisfactory properties, the most commonly used are carbonates, talc, aluminosilicates and hollow glass microspheres [10,34–37].

The present work aims to deeply investigate the effect of different dispersing agents on the interactions between the polymer matrix and hollow glass microspheres in fillers for nautical applications. In particular, both ionic (i.e., based on soy lecithin and on diamine dioleate) and steric dispersants (i.e., based on hyperbranched polyester and phosphite titanate) are tested by evaluating the rheological, mechanical and morphological properties of the prepared fillers. To the best of our knowledge, this is the first time that the interactions between the extenders and the matrix are carefully investigated and, remarkably, the rheological results are discussed by taking into account the Payne effect and providing a new perspective in understanding the behavior of such products.

2. Materials and Methods

2.1. Materials

The formulations studied here were prepared ad hoc in order to underline the effect of different dispersing agents on the wettability of hollow glass microspheres. Tested samples consisted of an epoxy resin (component A), an anti-foam agent necessary to avoid foam formation, hollow glass microspheres used as extenders [38] and a proper dispersing agent. Rheological modifiers, solvents, other extenders and pigments were not used to simplify the studied formulations. Four samples, which were labeled from 1 to 4, were prepared by employing different dispersing agents. Sample 0 was without the dispersing agent and was employed as a reference. Table 1 summarizes the composition of the un-crosslinked samples. It should be noted that the additives used here were commercial products, and therefore, their specific compositions are not available; the nature of the selected additives is reported in Table 2.

In order to obtain crosslinked products for the mechanical and morphological investigations, a polyamide resin (B component) was added to each sample. The amount of B component was calculated as a function of the epoxy resin content in each sample (for samples 0 and 4, 40.5 and 40.6 $w/w\%$, respectively; for samples 1 and 2, 3 and 40 $w/w\%$, respectively).

Resins (i.e., A and B components), anti-foam agent, and hollow glass microspheres have been kindly provided by Boero Bartolomeo S.p.A. (Boero Bartolomeo S.p.A., Genova, Italy). Self-emulsifying soy lecithin has been supplied by Balestrini S.r.l (Balestrini S.r.l, Milan, Italy. N-tallow alkyl trimethylene diamine dioleate has been provided by Eurochemicals S.p.A. (Eurochemicals S.p.A., Cologno Monzese, Italy). Hyperbranched polyester has been supplied by BYK-Chemie GmbH (BYK-Chemie GmbH, Wesel, Germany). Tetra(2,2-diallyloxymethylene-1-butyl)bis(ditridecyl phosphite) titanate has been provided by Finco S.r.l (Finco S.r.l., Settimo Milanese, Italy).

Table 1. Summary of the un-crosslinked sample compositions expressed in $w/w\%$.

Label	Component A	Dispersing Agent	Microspheres	Anti-Foam Agent
Sample 0 (without dispersing agent)	81.2	0.0	18.3	0.5
Sample 1 (with additive 1)	80.0	1.5	18.0	0.5
Sample 2 (with additive 2)	80.0	1.5	18.0	0.5
Sample 3 (with additive 3)	80.0	1.5	18.0	0.5
Sample 4 (with additive 4)	81.0	0.1	18.4	0.5

Table 2. Chemical nature of the dispersing agents used.

Dispersing Agent	Nature	Label
Additive 1	Ionic dispersants	Self-emulsifying soy lecithin
Additive 2		N-tallow alkyl trimethylene diamine dioleate
Additive 3	Steric dispersants	Hyperbranched polyester
Additive 4		Tetra(2,2-diallyloxymethylene-1-butyl)bis(ditridecyl phosphite) titanate

2.2. Methods

The dispersion of microspheres and additives in the polymer matrix was carried out with a dissolver Dispermat LC30, 220V (Dispermat®, VMA-Getzmann GmbH, Columbia, MD, USA). The mixture temperature, speed and time during the dispersion were controlled in order to have a good dispersion. For instance, the temperature was maintained below 40 °C, and the dissolver speed was set at 150–200 rpm. For each sample, both the un-crosslinked and crosslinked products were studied.

The rheological measurements were performed on the un-crosslinked materials using an Anton Paar MCR 102 rheometer (Anton Paar GmbH, Graz, Austria), equipped with a 25-mm-diameter parallel plate geometry (PP25) and using a 1-mm gap. The rheometer was equipped with a Peltier heating system for the accurate control of the temperature. All measurements were set at 25.00 ± 0.01 °C.

To evaluate the viscoelastic properties in terms of the storage modulus (i.e., G', representing the storage and recovery energy in cyclic deformation), loss modulus (i.e., G'', representing the energy dissipated as heat) and complex modulus (i.e., G* = G''/G'), amplitude sweep tests (AS) with a deformation (γ) ranging from 0.02% up to 10% were performed at a fixed frequency of 1 Hz. The data were collected and analyzed using RheoCompass software (Anton Paar GmbH, Graz, Austria). Each sample was tested in triplicate to ensure result repeatability.

The mechanical and morphological characterizations were performed on the crosslinked products obtained by mixing the samples with a proper amount of polyamide. Three-point bending flexural tests were performed according to ASTM D790 standard through a dynamometer (Instron 3365, Norwood, MA, USA) at room temperature [39–41]. Measurements were performed in triplicate on the samples to ensure result repeatability. For the morphological characterization, a Zeiss Supra 40VP Scanning Electron Microscope (Carl Zeiss AG, Oberkochen, Germany) was used. The samples were thinly coated with gold and palladium (0.150 kÅ Au/Pd) using a physical vapor deposition instrument (Precision Etching Coating System, Model 682, Gatan Inc., Pleasanton, CA, USA) in order to obtain good conductivity. Manual image analysis was carried out on digitalized SEM images using the open-source ImageJ 1.51 software (National Institute of Health, Bethesda, MD, USA to measure the distance between the polymer matrix and the hollow glass microspheres.

3. Results

3.1. Rheological Measurements

The rheological behavior of fillers is an important indicator of material applicability, as well as of the interactions occurring between their constituent components [18,42–44]. In particular, amplitude sweep tests are widely accepted to provide useful insights regarding the dispersion of extenders in a polymer matrix. Figures 1 and 2 report the viscoelastic moduli (i.e., G' and G'') and the complex modulus (i.e., G*) of the tested samples (i.e., 0–4), respectively, together with those of the simple matrix (i.e., pure epoxy resin without additives and microspheres).

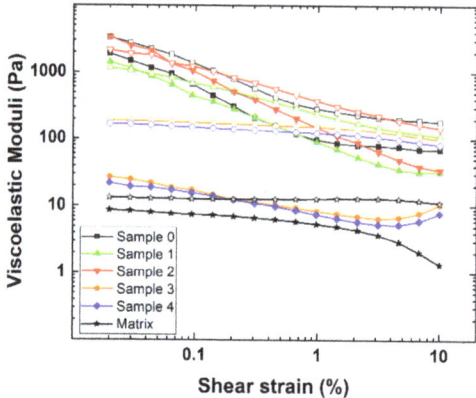

Figure 1. Dependence of the storage (filled symbols) and loss (empty symbols) moduli of Table 2. Dependence of the complex modulus of the tested samples upon the applied strain. The rheological response of the pure matrix is reported in the figure inset for comparison.

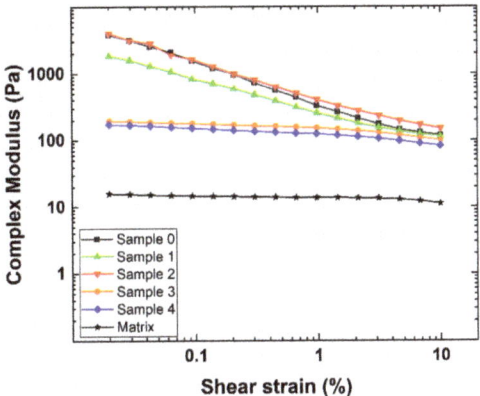

Figure 2. Dependence of the complex modulus of the tested samples upon the applied strain. The rheological response of the pure matrix is reported in the figure inset for comparison.

First of all, except for samples 1 and 2 at really low γ values, the loss modulus G″ always prevails over the elastic modulus G′ in the entire strain investigation range, therefore indicating that, in agreement with theory, the un-crosslinked materials show a prevalently viscous response [45]. Moreover, by comparing the matrix viscoelastic moduli with those of the samples (i.e., the polymer matrix with the added hollow glass microspheres), it can be noted that the presence of the extenders remarkably increases the material resistance, owing to their ability to both interact with the polymer matrix and form a secondary network [46]. Additionally, the tested samples can be clearly divided into two groups depending on their rheological behavior as a function of the applied strain. In more detail, samples 0, 1 and 2 present a high initial value of the moduli that rapidly decreases in around an order of magnitude with increasing γ. By contrast, samples 3 and 4 are characterized by lower initial values of G′, G″ and G* that slowly, and only slightly, decrease at the larger strain. Such findings can be explained with the Payne effect, which is a typical response of rubber-based composites loaded with extenders [47–49]. Additionally, the mechanism responsible for the Payne effect is still controversial and not completely understood. The most commonly accepted explanation is related to the secondary network (extender–extender) formed by the extenders within the polymer matrix. At small amplitudes, this structure is able to act as a reinforcement, whereas it gets progressively destroyed upon the application of a greater oscillatory strain, leading to a marked decrease of the material resistance to solicitations. The larger the Payne effect, the greater the extender-extender interactions are at the expenses of those between the extenders and the matrix [50–53]. The marked Payne effect shown by samples 1 and 2, similar to that observed for sample 0 (i.e., the reference sample without a dispersing agent), is indicative of the fact that hollow glass microspheres are not efficiently dispersed and can form agglomerates, thus indicating the low efficiency of the ionic dispersing agents. On the other hand, the small Payne effect depicted for samples 3 and 4 clearly suggests the capability and proficiency of the steric dispersing agents in homogeneously dispersing the microspheres used, therefore promoting the extender-matrix interactions. To quantitatively evaluate the Payne effect of the tested samples, a widely employed approach consists in considering the Payne amplitude, ΔG^*, as the difference between the complex modulus, G_0^*, at very low strain values (i.e., 0.02%), and the complex modulus, G_∞^*, at high strain values (i.e., 10%). The obtained results are summarized in Table 3.

In agreement with the rheological response reported in Figures 1 and 2, and taking into account the above discussion, the Payne amplitude evaluation clearly demonstrates that additives 3 and 4 perform well in dispersing hollow glass microspheres within the polymer matrix and are able to provide a response similar to that of the polymer matrix. Conversely, additives 1 and 2 offer a negligible, or even negative, dispersing effect and do not provide any significant difference compared to sample 0 (i.e., the reference sample). Owing to the different nature of the dispersing agents used, such findings provide the first evidence that steric surfactants are much more efficient in the investigated fillers by exploiting their dispersing action due to steric hindrance [54,55].

Table 3. The complex moduli, G_0^* (at very low strain) and G_∞^* (at high strain), and Payne amplitude (ΔG^*) values for the tested polymer matrix and samples.

Sample.	G_0^* (Pa)	G_∞^* (Pa)	ΔG^* (Pa)
Matrix	15 ± 1	11 ± 0.5	4.8 ± 0.9
Sample 0	3812 ± 11	119 ± 1	3693 ± 11
Sample 1	1733 ± 70	104 ± 9	1629 ± 61
Sample 2	3951 ± 4	148 ± 1	3803 ± 4
Sample 3	193 ± 4	101 ± 3	91 ± 1
Sample 4	169 ± 2	83 ± 4	87 ± 3

3.2. Mechanical Tests

Compared to rheological tests performed on un-crosslinked formulations, the mechanical response of solid crosslinked materials can be employed to evaluate the reinforcing effect of extenders in nautical fillers. Indeed, the predominance of extender-extender interactions results in poor mechanical performance due to the impossibility to efficiently transfer an applied stress between the material components with the consequent formation of weak spots; conversely, a good compatibility between the extenders and the matrix allows obtaining a much more homogeneous and performing material with an enhanced response compared to the pure polymer [56]. Here, mechanical bending tests were performed on samples with a thickness of 0.8 cm, a width of 2.0 cm and a length of 20 cm. In particular, the Young modulus (E_b), the bending strength (σ_b) and the deformation at break (ε_b) were calculated, with the results summarized in Table 4.

Table 4. Summary of the samples' mechanical and morphological properties.

Sample	E_b (MPa)	σ_b (MPa)	ε_b (%)	Extender–Matrix Distance (μm)
Sample 0	1454 ± 90	24.9 ± 1.8	1.80 ± 0.17	0.254 ± 0.047
Sample 1	1444 ± 31	17.7 ± 0.9	1.34 ± 0.11	0.235 ± 0.023
Sample 2	1490 ± 53	18.0 ± 0.7	1.26 ± 0.07	0.249 ± 0.051
Sample 3	1866 ± 32	24.9 ± 1.3	1.38 ± 0.13	-
Sample 4	1857 ± 59	24.4 ± 1.5	1.36 ± 0.14	-

In agreement with the rheological results, two sample groups can be clearly individuated on the basis of their mechanical response. In detail, using ionic dispersants (i.e., samples 1 and 2) was found to decrease both the break strength and elongation of the system with values of around 18 MPa and 1.3%, respectively, and no differences were observed for the elastic modulus (i.e., ~1500 MPa) with respect to sample 0. The observed break strength decrement is probably ascribable to a slight plasticizing action caused by the ionic dispersants [33,57,58], also taking into account that the studied hollow glass microspheres offer a modest reinforcement effect compared to other filler types. By contrast, the steric dispersants (i.e., samples 3 and 4) induced a considerable increment of the system elastic modulus, without affecting the break strength and only slightly reducing the filler deformability [59–61]. Compared to the rheological results, such findings indicate that steric dispersants are able to provide a much more marked compatibilization effect for the studied system (i.e., hollow glass microspheres embedded in an epoxy resin matrix), as well as the fact that ionic dispersants are almost completely ineffective and can even to some extent worsen the filler mechanical response.

3.3. Morphological Characterization

A simple and fast approach to qualitatively estimate extender-matrix interactions in crosslinked fillers relies on evaluating the distance between the two components via morphological characterization. SEM images for samples 0, 1 and 2 and for samples 3 and 4 are reported in Figures 3 and 4, respectively, with the extender-matrix distance (d) summarized in Table 4.

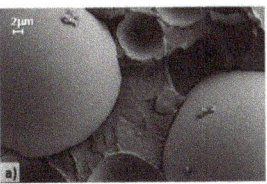 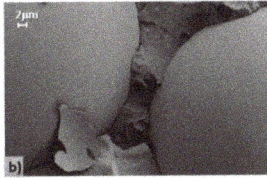 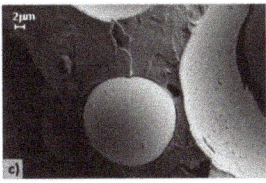

Figure 3. SEM images of samples (**a**) 0, (**b**) 1, and (**c**) 2.

 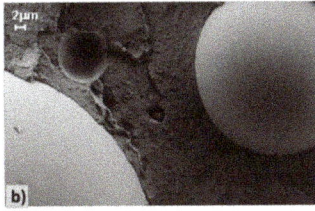

Figure 4. SEM images of samples (**a**) 3 and (**b**) 4.

As clearly shown, the glass microsphere wettability strongly depends on the employed dispersing agents and reflects the rheological and mechanical results. Sample 0 (Figure 3a) is characterized by a well-defined empty region between the extenders and the polymers (d = 0.254 µm), thus suggesting the complete incompatibility between these components. Similarly, samples 1 (Figure 3b) and 2 (Figure 3c) present the same morphology of the reference sample, in addition to the presence of the dispersing agents (d = 0.235 and 0.249 µm, respectively), which consequently can be considered totally unable to create effective interactions between the extenders and the matrix. Conversely, samples 3 (Figure 4a) and 4 (Figure 4b) are characterized by a different morphology, where a neat interface between the components cannot be clearly depicted, thus proving their good compatibility and the existence of a continuous composite structure with enhanced performance. Note that it was not possible to calculate the extender-matrix distance for the last two samples.

To better visualize the described phenomenon, Figure 5 shows the SEM images at high magnification of sample 2 (Figure 5a), which is characterized by the presence of an ineffective dispersing agent, and sample 3 (Figure 5b), which is instead characterized by the presence of an efficient dispersing agent. As clearly highlighted by the white arrows, a neat extender-matrix interface can be observed in Figure 5a. By contrast, an almost continuum medium with a slightly detectable interface is depicted in Figure 5b.

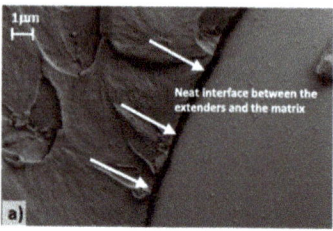 

Figure 5. High-magnification SEM images of samples (**a**) 2 and (**b**) 3.

4. Conclusions

In the present work, the effect of dispersing agents with a different action mode (i.e., ionic or steric) on the interaction between hollow glass microspheres, which were used as extenders, and an epoxy-polyamide resin, which represented the typical polymer matrix used in nautical fillers, was investigated. The rheological behavior of the studied samples clearly indicated that the performance of the steric surfactants was much more enhanced in reducing the extender-extender interactions, compared to the ionic ones. In more detail, the Payne effect, which consists of a marked decrease of the material viscoelastic moduli G' and G" upon the application of an oscillatory shear, was found to be much more evident in the presence of the ionic additives, thus indicating their poor efficiency in homogeneously dispersing the microspheres. Additionally, bending tests proved that the steric dispersants improved the mechanical resistance of the fillers, owing to their capability to form a continuous complex structure with an enhanced response. Remarkably, the sample morphological investigation allowed for the clear visualization of the effect of the different dispersing agents on the wettability of the glass extenders; in particular, whereas a neat interfacial region could be detected for the ionic surfactants, the steric ones led to a much greater adhesion of the two components, reflecting the previous findings. However, further experiments are clearly needed to fully understand the described phenomenon. This work represents the first scientific report concerning the evaluation of the effect of different dispersing agents on the performance of fillers in the nautical field.

Author Contributions: Conceptualization, S.V. (Silvia Vita), R.R. and S.V. (Silvia Vicini); methodology, S.V. (Silvia Vita), R.R. and A.D.; validation, R.R., S.V. (Silvia Vicini) and M.C.; formal analysis, S.V. (Silvia Vita) and R.R.; investigation, S.V. (Silvia Vita) and P.B.; resources, S.V. (Silvia Vicini) and E.P.; data curation, S.V. (Silvia Vita), R.R. and A.D.; writing—original draft preparation, S.V. (Silvia Vita) and A.D.; writing—review and editing, S.V. (Silvia Vita), R.R., A.D., S.V. (Silvia Vicini) and M.C.; supervision, S.V. (Silvia Vicini) and M.C. All authors have read and agreed to the published version of the manuscript.

Funding: This research received no external funding.

Conflicts of Interest: The authors declare no conflicts of interest.

References

1. Wallström, E.; Jespersen, H.T.; Schaumburg, K. A new concept for anti-fouling paint for Yachts. In *Proceedings of the Progress in Organic Coatings*; Elsevier: Amsterdam, The Netherlands, 2011; Volume 72, pp. 109–114.
2. Böß, V.; Denkena, B.; Dittrich, M.A.; Kenneweg, R. Mathematical description of aesthetic criteria for process planning and quality control of luxury yachts. In *Proceedings of the Procedia CIRP*; Elsevier B.V.: Amsterdam, The Netherlands, 2019; Volume 79, pp. 478–483.
3. Armendáriz-Ontiveros, M.M.; Fimbres Weihs, G.A.; de los Santos Villalobos, S.; Salinas-Rodriguez, S.G. Biofouling of FeNP-Coated SWRO Membranes with Bacteria Isolated after Pre-Treatment in the Sea of Cortez. *Coatings* **2019**, *9*, 462. [CrossRef]
4. Akuzov, D.; Franca, L.; Grunwald, I.; Vladkova, T. Sharply Reduced Biofilm Formation from Cobetia marina and in Black Sea Water on Modified Siloxane Coatings. *Coatings* **2018**, *8*, 136. [CrossRef]

5. Delucchi, M.; Castellano, M.; Vicini, S.; Vita, S.; Finocchio, E.; Ricotti, R.; Cerisola, G. A methodological approach for monitoring the curing process of fairing compounds based on epoxy resins. *Prog. Org. Coat.* **2018**, *123*, 20–26. [CrossRef]
6. Boote, D.; Vergassola, G.; Giannarelli, D.; Ricotti, R. Thermal load effects on side plates of superyachts. *Mar. Struct.* **2017**, *56*, 39–68. [CrossRef]
7. Strong, A.B. *Fundamentals of Composites Manufacturing: Materials, Methods and Applications*; Society of Manufacturing Engineers: Dearborn, MI, USA, 2008; ISBN 9781613449677.
8. Delucchi, M.; Finocchio, E.; Castellano, M.; Vicini, S.; Vita, S.; Cerisola, G.; Ricotti, R. Application of DSC and FTIR techniques for monitoring the curing process of epoxy fillers used for yacht application. *Metall. Ital.* **2017**, *109*, 107–110.
9. Sharma, S.; Wetzel, K.K. Process Development Issues of Glass—Carbon Hybrid-reinforced Polymer Composite Wind Turbine Blades. *J. Compos. Mater.* **2010**, *44*, 437–456. [CrossRef]
10. Delucchi, M.; Ricotti, R.; Cerisola, G. Influence of micro- and nano-fillers on chemico-physical properties of epoxy-based materials. In Proceedings of the Progress in Organic Coatings; Elsevier: Amsterdam, The Netherlands, 2011; Volume 72, pp. 58–64.
11. Donnelly, B.; Bedwell, I.; Dimas, J.; Scardino, A.; Tang, Y.; Sammut, K. Effects of Various Antifouling Coatings and Fouling on Marine Sonar Performance. *Polymers (Basel)* **2019**, *11*, 663. [CrossRef]
12. Santos, T.; Nunes, L.; Faria, P. Production of eco-efficient earth-based plasters: Influence of composition on physical performance and bio-susceptibility. *J. Clean. Prod.* **2017**, *167*, 55–67. [CrossRef]
13. Bochen, J.; Labus, M. Study on physical and chemical properties of external lime-sand plasters of some historical buildings. *Constr. Build. Mater.* **2013**, *45*, 11–19. [CrossRef]
14. Graham-Jones, J.; Summerscales, J. *Marine Applications of Advanced Fibre-Reinforced Composites*; Woodhead Publishing: Cambridge, UK, 2015; ISBN 9780081002001.
15. Melià, P.; Ruggieri, G.; Sabbadini, S.; Dotelli, G. Environmental impacts of natural and conventional building materials: A case study on earth plasters. *J. Clean. Prod.* **2014**, *80*, 179–186. [CrossRef]
16. Micó-Vicent, B.; Jordán, J.; Perales, E.; Martínez-Verdú, F.M.; Cases, F. Finding the Additives Incorporation Moment in Hybrid Natural Pigments Synthesis to Improve Bioresin Properties. *Coatings* **2019**, *9*, 34. [CrossRef]
17. Viesca, J.-L.; Anand, M.; Blanco, D.; Fernández-González, A.; García, A.; Hadfield, M. Tribological Behaviour of PVD Coatings Lubricated with a FAP− Anion-Based Ionic Liquid Used as an Additive. *Lubricants* **2016**, *4*, 8. [CrossRef]
18. Gutiérrez-González, S.; Alonso, M.M.; Gadea, J.; Rodríguez, A.; Calderón, V. Rheological behaviour of gypsum plaster pastes with polyamide powder wastes. *Constr. Build. Mater.* **2013**, *38*, 407–412. [CrossRef]
19. Patton, T.C. *Paint Flow and Pigment Dispersion: A Rheological Approach to Coating and Ink*; Wiley: Hoboken, NJ, USA, 1966; Volume 68, ISBN 978-0-471-03272-4.
20. Tracton, A.A. *Coatings Materials and Surface Coatings*; CRC Press: Boca Raton, FL, USA, 2006.
21. Shukla, S.; Seal, S. Thermodynamic tetragonal phase stability in sol-gel derived nanodomains of pure zirconia. *J. Phys. Chem. B* **2004**, *108*, 3395–3399. [CrossRef]
22. Fujibayashi, T.; Okubo, M. Preparation and thermodynamic stability of micron-sized, monodisperse composite polymer particles of disc-like shapes by seeded dispersion polymerization. *Langmuir* **2007**, *23*, 7958–7962. [CrossRef]
23. Jaglinski, T.; Kochmann, D.; Stone, D.; Lakes, R.S. Composite materials with viscoelastic stiffness greater than diamond. *Science* **2007**, *315*, 620–622. [CrossRef]
24. Lova, P.; Giusto, P.; Di Stasio, F.; Manfredi, G.; Paternò, G.M.; Cortecchia, D.; Soci, C.; Comoretto, D. All-polymer methylammonium lead iodide perovskite microcavities. *Nanoscale* **2019**, *11*, 8978–8983. [CrossRef]
25. Tsai, Y.T.; Chiou, J.Y.; Liao, C.Y.; Chen, P.Y.; Tung, S.H.; Lin, J.J. Organically modified clays as rheology modifiers and dispersing agents for epoxy packing of white LED. *Compos. Sci. Technol.* **2016**, *132*, 9–15. [CrossRef]
26. Boccalero, G.; Jean-Mistral, C.; Castellano, M.; Boragno, C. Soft, hyper-elastic and highly-stable silicone-organo-clay dielectric elastomer for energy harvesting and actuation applications. *Compos. Part B Eng.* **2018**, *146*, 13–19. [CrossRef]

27. Castellano, M.; Turturro, A.; Riani, P.; Montanari, T.; Finocchio, E.; Ramis, G.; Busca, G. Bulk and surface properties of commercial kaolins. *Appl. Clay Sci.* **2010**, *48*, 446–454. [CrossRef]
28. Holmberg, K. Natural surfactants. *Curr. Opin. Colloid Interface Sci.* **2001**, *6*, 148–159. [CrossRef]
29. Shinoda, W.; Devane, R.; Klein, M.L. Coarse-grained molecular modeling of non-ionic surfactant self-assembly. *Soft Matter* **2008**, *4*, 2454–2462. [CrossRef]
30. Solè, I.; Maestro, A.; González, C.; Solans, C.; Gutiérrez, J.M. Optimization of nano-emulsion preparation by low-energy methods in an ionic surfactant system. *Langmuir* **2006**, *22*, 8326–8332. [CrossRef] [PubMed]
31. Yin, J.; Migas, D.B.; Panahandeh-Fard, M.; Chen, S.; Wang, Z.; Lova, P.; Soci, C. Charge redistribution at GaAs/P3HT heterointerfaces with different surface polarity. *J. Phys. Chem. Lett.* **2013**, *4*, 3303–3309. [CrossRef]
32. Castellano, M.; Alloisio, M.; Darawish, R.; Dodero, A.; Vicini, S. Electrospun composite mats of alginate with embedded silver nanoparticles. *J. Therm. Anal. Calorim.* **2019**, *137*, 767–778. [CrossRef]
33. Shamsuri, A.A.; Md. Jamil, S.N.A. Compatibilization Effect of Ionic Liquid-Based Surfactants on Physicochemical Properties of PBS/Rice Starch Blends: An Initial Study. *Materials (Basel)* **2020**, *13*, 1885. [CrossRef]
34. Bertora, A.; Castellano, M.; Marsano, E.; Alessi, M.; Conzatti, L.; Stagnaro, P.; Colucci, G.; Priola, A.; Turturro, A. A new modifier for silica in reinforcing SBR elastomers for the tyre industry. *Macromol. Mater. Eng.* **2011**, *296*, 455–464. [CrossRef]
35. Castellano, M.; Marsano, E.; Turturro, A.; Conzatti, L.; Busca, G. Dependence of surface properties of silylated silica on the length of silane arms. *Adsorption* **2012**, *18*, 307–320. [CrossRef]
36. Tarsi, G.; Caputo, P.; Porto, M.; Sangiorgi, C. A Study of Rubber-REOB Extender to Produce Sustainable Modified Bitumens. *Appl. Sci.* **2020**, *10*, 1204. [CrossRef]
37. Dörr, D.; Standau, T.; Murillo Castellón, S.; Bonten, C.; Altstädt, V. Rheology in the Presence of Carbon Dioxide (CO_2) to Study the Melt Behavior of Chemically Modified Polylactide (PLA). *Polymers (Basel)* **2020**, *12*, 1108. [CrossRef]
38. Shira, S.; Buller, C. *Hollow Glass Microspheres for Plastics, Elastomers, and Adhesives Compounds*; Elsevier: Amsterdam, The Netherlands, 2015; pp. 241–271. [CrossRef]
39. ASTM D790 - 17 Standard Test Methods for Flexural Properties of Unreinforced and Reinforced Plastics and Electrical Insulating Materials. Available online: https://www.astm.org/Standards/D790 (accessed on 8 June 2020).
40. Glória, G.O.; Teles, M.C.A.; Neves, A.C.C.; Vieira, C.M.F.; Lopes, F.P.D.; de Gomes, M.A.; Margem, F.M.; Monteiro, S.N. Bending test in epoxy composites reinforced with continuous and aligned PALF fibers. *J. Mater. Res. Technol.* **2017**, *6*, 411–416. [CrossRef]
41. Linhares, F.N.; Gabriel, C.F.S.; de Sousa, A.M.F.; Nunes, R.C.R. Mechanical and rheological properties of nitrile rubber/fluoromica composites. *Appl. Clay Sci.* **2018**, *162*, 165–174. [CrossRef]
42. Senff, L.; Ascensão, G.; Ferreira, V.M.; Seabra, M.P.; Labrincha, J.A. Development of multifunctional plaster using nano-TiO_2 and distinct particle size cellulose fibers. *Energy Build.* **2018**, *158*, 721–735. [CrossRef]
43. Ochoa, R.E.; Gutiérrez, C.A.; López-Cuevas, J.; Rendón, J.; Rodríguez-Galicia, J.L.; Cruz-Álvarez, J. Effect of Water/Plaster Ratio on Preparing Molds for Slip Casting of Sanitaryware; Rheology of the Initial Plaster Slurry, Microstructure and Mold Properties. *Trans. Indian Ceram. Soc.* **2018**, *77*, 84–89. [CrossRef]
44. Dodero, A.; Alloisio, M.; Vicini, S.; Castellano, M. Preparation of composite alginate-based electrospun membranes loaded with ZnO nanoparticles. *Carbohydr. Polym.* **2020**, *227*, 115371. [CrossRef]
45. Chaudhary, A.K.; Jayaraman, K. Extrusion of linear polypropylene-clay nanocomposite foams. *Polym. Eng. Sci.* **2011**, *51*, 1749–1756. [CrossRef]
46. Brunengo, E.; Castellano, M.; Conzatti, L.; Canu, G.; Buscaglia, V.; Stagnaro, P. PVDF-based composites containing PZT particles: How processing affects the final properties. *J. Appl. Polym. Sci.* **2020**, *137*, 48871. [CrossRef]
47. Payne, A.R. The dynamic properties of carbon black loaded natural rubber vulcanizates. Part II. *J. Appl. Polym. Sci.* **1962**, *6*, 368–372. [CrossRef]
48. Hayeemasae, N.; Sensem, Z.; Surya, I.; Sahakaro, K.; Ismail, H. Synergistic Effect of Maleated Natural Rubber and Modified Palm Stearin as Dual Compatibilizers in Composites based on Natural Rubber and Halloysite Nanotubes. *Polymers (Basel)* **2020**, *12*, 766. [CrossRef]

49. Srivastava, S.; Mishra, Y. Nanocarbon Reinforced Rubber Nanocomposites: Detailed Insights about Mechanical, Dynamical Mechanical Properties, Payne, and Mullin Effects. *Nanomaterials* **2018**, *8*, 945. [CrossRef] [PubMed]
50. Ramier, J.; Gauthier, C.; Chazeau, L.; Stelandre, L.; Guy, L. Payne effect in silica-filled styrene–butadiene rubber: Influence of surface treatment. *J. Polym. Sci. Part B Polym. Phys.* **2007**, *45*, 286–298. [CrossRef]
51. Castellano, M.; Turturro, A.; Marsano, E.; Conzatti, L.; Vicini, S. Hydrophobation of silica surface by silylation with new organo-silanes bearing a polybutadiene oligomer tail. *Polym. Compos.* **2014**, *35*, 1603–1613. [CrossRef]
52. Hentschke, R. The payne effect revisited. *Express Polym. Lett.* **2017**, *11*, 278–292. [CrossRef]
53. Xu, H.; Fan, T.; Ye, N.; Wu, W.; Huang, D.; Wang, D.; Wang, Z.; Zhang, L. Plasticization Effect of Bio-Based Plasticizers from Soybean Oil for Tire Tread Rubber. *Polymers (Basel)* **2020**, *12*, 623. [CrossRef]
54. Aranguren, M.I.; Mora, E.; DeGroot, J.V.; Macosko, C.W. Effect of reinforcing fillers on the rheology of polymer melts. *J. Rheol.* **1992**, *36*, 1165–1182. [CrossRef]
55. Cassagnau, P. Melt rheology of organoclay and fumed silica nanocomposites. *Polymer* **2008**, *49*, 2183–2196. [CrossRef]
56. Vallittu, P.K. High-aspect ratio fillers: Fiber-reinforced composites and their anisotropic properties. *Dent. Mater.* **2015**, *31*, 1–7. [CrossRef]
57. Dybowska-Sarapuk, L.; Kielbasinski, K.; Arazna, A.; Futera, K.; Skalski, A.; Janczak, D.; Sloma, M.; Jakubowska, M. Efficient Inkjet Printing of Graphene-Based Elements: Influence of Dispersing Agent on Ink Viscosity. *Nanomaterials* **2018**, *8*, 602. [CrossRef]
58. Ng, S.; Justnes, H. Influence of dispersing agents on the rheology and early heat of hydration of blended cements with high loading of calcined marl. *Cem. Concr. Compos.* **2015**, *60*, 123–134. [CrossRef]
59. Zhan, Z.; He, H.; Zhu, Z.; Xue, B.; Wang, G.; Chen, M.; Xiong, C. Blends of rABS and SEBS: Influence of In-Situ Compatibilization on the Mechanical Properties. *Materials (Basel)* **2019**, *12*, 2352. [CrossRef] [PubMed]
60. Chen, J.; Wang, Y.; Gu, C.; Liu, J.; Liu, Y.; Li, M.; Lu, Y. Enhancement of the Mechanical Properties of Basalt Fiber-Wood-Plastic Composites via Maleic Anhydride Grafted High-Density Polyethylene (MAPE) Addition. *Materials (Basel)* **2013**, *6*, 2483–2496. [CrossRef] [PubMed]
61. Quitadamo, A.; Massardier, V.; Santulli, C.; Valente, M. Optimization of Thermoplastic Blend Matrix HDPE/PLA with Different Types and Levels of Coupling Agents. *Materials (Basel)* **2018**, *11*, 2527. [CrossRef] [PubMed]

© 2020 by the authors. Licensee MDPI, Basel, Switzerland. This article is an open access article distributed under the terms and conditions of the Creative Commons Attribution (CC BY) license (http://creativecommons.org/licenses/by/4.0/).

Article

Performance of Nano- and Microcalcium Carbonate in Uncrosslinked Natural Rubber Composites: New Results of Structure–Properties Relationship

Nantikan Phuhiangpa [1], Worachai Ponloa [1], Saree Phongphanphanee [1] and Wirasak Smitthipong [1,2,3,*]

1. Specialized Center of Rubber and Polymer Materials in Agriculture and Industry (RPM), Department of Materials Science, Faculty of Science, Kasetsart University, Bangkok 10900, Thailand; wernnantikan@gmail.com (N.P.); worachai.p@ku.th (W.P.); fscisrph@ku.ac.th (S.P.)
2. Office of Research Integration on Target-Based Natural Rubber, National Research Council of Thailand (NRCT), Bangkok 10900, Thailand
3. Office of Natural Rubber Research Program, Thailand Science Research and Innovation (TSRI), Bangkok 10400, Thailand
* Correspondence: fsciwssm@ku.ac.th

Received: 11 August 2020; Accepted: 1 September 2020; Published: 3 September 2020

Abstract: Calcium carbonate ($CaCO_3$) is one of the most important inorganic powders and is widely used as filler in order to reduce costs in the rubber industry. Nanocalcium carbonate reduces costs and acts as a semireinforcing filler that improves the mechanical properties of rubber composites. The objective of this study was to investigate the effect of nano-$CaCO_3$ (NCC) and micro-$CaCO_3$ (MCC) on the properties of natural rubber composites, in particular, new results of structure–properties relationship. The effects of NCC/MCC on the properties of rubber composites, such as Mooney viscosity, bound rubber, Mullins effect, and Payne effect, were investigated. The result of the Mullins effect of rubber composites filled with NCC was in good agreement with the results of Mooney viscosity and bound rubber, with higher Mooney viscosity and bound rubber leading to higher stress to pull the rubber composites. The Payne effect showed that the value of different storage moduli ($\Delta G'$) of rubber composites filled with 25 parts per hundred rubber (phr) NCC was the lowest due to weaker filler network, while the rubber supplemented with 100 phr NCC had more significant $\Delta G'$ values with increase in strain. The results of rubber composites filled with MCC showed the same tendency as those of rubber composites filled with NCC. However, the effect of specific surface area of NCC on the properties of rubber composites was more pronounced than those of rubber composites filled with MCC. Finite element analysis of the mechanical property of rubber composites was in good agreement with the result from the experiment. The master curves of time–temperature superposition presented lower free volume in the composites for higher loading of filler, which would require more relaxation time of rubber molecules. This type of nanocalcium carbonate material can be applied to tailor the properties and processability of rubber products.

Keywords: nanocalcium carbonate; natural rubber nanocomposite; Mullins effect; stress relaxation; Payne effect

1. Introduction

Natural rubber (NR) is a major agricultural product that is widely used in the rubber industry for the production of tile floor, tires, gloves, pillows and mattresses, medical products, etc. NR can be obtained from *Hevea brasiliensis*, which consists of rubber (*cis*-1,4-polyisoprene) and nonrubber components (proteins, phospholipids, sugars, salts, etc.) [1–3]. To date, there is no other synthetic material that can replace natural rubber from plants [4]. NR is often reinforced by incorporation of filler to improve its mechanical properties, namely, modulus, hardness, tensile strength, abrasion resistance, and tear resistance [5].

Recently, fillers have been widely used in the rubber industry for many purposes, such as improvement of mechanical properties, efficient production, and reduction in cost of rubber products [6–11]. Generally, there are three main groups of fillers: reinforcing fillers, semireinforcing fillers, and nonreinforcing fillers. The efficiency of the reinforcing filler depends on several factors, such as particle size, surface area, and the shape of filler [12–14]. Among commercial fillers, carbon black and silica (SiO_2) are the most important reinforcing fillers. They are added to improve the mechanical properties of rubber compounds. However, there are some fillers that can be used as non- or semireinforcing fillers (such as calcium carbonate, clay, talc, etc.) for either reducing cost or improving mechanical properties [15].

Calcium carbonate ($CaCO_3$) is one of the most important inorganic powders and is widely used as filler in paints, plastics, and the rubber industry in order to reduce the material cost [16]. Calcium carbonate is considered as a filler for rubbers. Its surface property, controlled shape, and small particle size affect the processing and properties of rubber composites. Calcium carbonate can be coated by hydrophobic molecules in order to better interact with hydrophobic rubber [17]. Based on modern technology, the particle size of $CaCO_3$ can be reduced to the nanoscale [18]. Nanocalcium carbonate (NCC) can be added to rubber, and the mechanical properties of the nanocomposites increases with increasing amount of NCC [19]. Moreover, the dynamic characteristics of NCC added to crosslinked NR has been studied, and the results showed that the shape of NCC affects the Mullins and Payne effects of the rubber composite [20]. NCC can be used as fillers not only to decrease the cost of materials but also to increase the mechanical properties of the crosslinked rubber composites [21].

However, there is no known research on the structure–properties relationship of nanocalcium carbonate in uncrosslinked NR nanocomposites compared to microcalcium carbonate in uncrosslinked NR composites, which can be beneficial for potential applications of nanocalcium carbonate in the rubber industry. Thus, the main aim of this work was to investigate the effect of both nano-$CaCO_3$ (NCC) and micro-$CaCO_3$ (MCC) at a wide range of filler loading on the properties of uncrosslinked rubber composites, in particular, new results of structure–properties relationship.

2. Materials and Methods

2.1. Materials

The main material were nanocalcium carbonate (particle size around 80 nm, Sand and Soil Industry Co., Ltd., Bangkok, Thailand), microcalcium carbonate (particle size around 1.30 µm, Sand and Soil Industry Co., Ltd., Bangkok, Thailand), natural rubber (STR 5L, Rubber Authority of Thailand, Bangkok, Thailand), and toluene (AR grade, RCI Labscan Limited, Bangkok, Thailand).

2.2. Preparation of Composites

Natural rubber, in total weight of rubber for 500 g of each formulation, was first added into a two-roll mill and masticated at 70 °C for 5 min. After that, the amount of either NCC or MCC was added into the NR according to Table 1 and then mixed together at 70 °C for 15 min. Finally, all samples were cut into the form of testing sheets.

Table 1. Formulations of rubber composites.

Ingredients	Parts Per Hundred of Rubber (phr) for Each Formulation								
	NR	NR/NCC	NR/NCC	NR/NCC	NR/NCC	NR/MCC	NR/MCC	NR/MCC	NR/MCC
NR	100	100	100	100	100	100	100	100	100
NCC	-	25	50	75	100	-	-	-	-
MCC	-	-	-	-	-	25	50	75	100

2.3. Characterizations

For scanning electron microscopy analysis (SEM; FEI, Quanta 450 FEI, Eindhoven, Netherlands), rubber samples were cut into small pieces and coated with gold in a sputter coater (Polaron Range SC7620, Quorum Technologies Ltd., Kent, UK) for the morphology analysis.

The Mooney viscosity of rubber composites was determined by a Mooney viscometer (viscTECH+, Techpro, Columbia city, IN, USA); the weight of the sample was 12 g. The Mooney viscometer consists of a rotating disc imbedded in a rubber specimen contained within a sealed, pressurized, and heated cavity. The rubber sample was heated at 100 °C for 1 min before starting the motor. After that, the rubber sample was continuously measured for 4 min using the torque required to keep the rotor rotating at a constant rate as a function of time for reading the Mooney viscosity, which was recorded as torque in newton meter (Nm).

Bound rubber was considered a quantitative measure of the filler surface activity and the rubber–filler interaction. Bound rubber was determined by immersing 1 g of rubber composite in 100 mL of toluene solvent at room temperature for 7 days. After dissolution, a piece of rubber that is insoluble in toluene was filtered, weighed, and then calculated with respect to the original sample weight [22].

The Mullins effect of the rubber composites was determined by a universal testing machine (AGS-X,20 N, Shimadzu, Tokyo, Japan); the rubber sample was cut into dumbbell shape. The condition of test in the strain axis was varied 1–120% in cycle mode, and then pulling–releasing of the sample was carried out three times.

Stress relaxation of the rubber composites was determined by dynamic mechanical analysis (DMA1, Mettler Toledo, Columbus, OH, USA). The rubber was cut into samples of 2 mm width, 4 mm length, and 1.5 mm thickness and tested at a temperature of 30 °C with strain range of 10% for 600 s. A curve measured on a DMA-type module using segments of the stress relaxation type can be displayed in a stress relaxation diagram as a stress–time curve. The stress–time curve presented the actual force measurement signal and the cross-sectional area of the sample as a function of time.

The Payne effect of either NCC- or MCC-supplemented rubber composites was analyzed using a rubber processing analyzer (RPA 2000, Alpha Technologies, Hudson, OH, USA) under the following conditions: temperature 60 °C, frequency 1 Hz, strain range 1–100%. The value of different storage moduli ($\Delta G'$) means $G'_{max} - G'_{min}$.

Viscoelastic properties of the rubber composites were determined by dynamic mechanical analysis (DMA1, Mettler Toledo, Columbus, OH, USA) based on Williams–Landel–Ferry (WLF) analysis; the rubber composite was cut into samples of 2 mm width, 4 mm length, and 1.5 mm thickness and tested at a temperature range of −80 to 50 °C and frequency of 1–100 Hz. The time–temperature superposition principle was used to establish a master curve of the storage modulus (E′) as a function of reduced frequency at a reference temperature T_{ref} of 298 K [23]. Shift factors (a_T) for the establishment of master curves were determined according to the WLF equation, where C_1 and C_2 are constants depending on the nature of the elastomer and the reference temperature [24].

2.4. Finite Element Analysis

The material parameters for the rubber composites were determined by incompressible isotropic hyperelastic strain energy models. In this work, we applied the reduced polynomial form of strain energy potential from order 3 (or Yeoh model) to 6. To get the parameters of the models, the finite element method (FEM) and curve fitting analysis were carried out using ABAQUS on uniaxial tension of the rubber composites. For the hyperelastic materials, the strain energy function of incompressible is the function of strain invariants, I [25]:

$$W = W(I_1, I_2, I_3) \tag{1}$$

The strain invariants can be written as follows:

$$I_1 = \lambda_1^2 + \lambda_2^2 + \lambda_3^2 \tag{2}$$

$$I_2 = \lambda_1^2\lambda_2^2 + \lambda_2^2\lambda_3^2 + \lambda_3^2\lambda_1^2 \tag{3}$$

$$I_3 = \lambda_1^2\lambda_2^2\lambda_3^2 \tag{4}$$

where λ_1, λ_2, and λ_3 are the principal stretches. The strain energy function of reduced polynomial for incompressible rubber can be written as follows:

$$W = \sum_{i=1}^{N} C_{i0}(I_1 - 3)^i \tag{5}$$

where C_{i0} is the temperature-dependent parameter for materials. The parameter C_{10} is related to the initial shear modulus, μ_0, by $2C_{10} = \mu_0$.

3. Results and Discussion

3.1. Morphological Properties

The SEM images (Figure 1) of MCC and NCC showed spherical shape for both types of calcium carbonates. These results are in good agreement with previous works [20,26]. It was observed that the MCC particles had larger particles compared to NCC at the same magnification. Regarding the supplier certificate, MCC had around 1.30 µm particle size compared to around 80 nm for NCC. Besides the smaller primary particle size of NCC, it could be seen that the nanoparticles were grouped together into aggregates. As each aggregate still had attraction forces from the specific surface area, those groups could bundle further to form larger agglomerates. Nevertheless, the primary particle size of NCC was, on average, an order of magnitude smaller than MCC, even though both had the same spherical shape.

(a)   (b)

Figure 1. The SEM images of (a) micro-CaCO$_3$ (MCC) and (b) nano-CaCO$_3$ (NCC) in spherical shape.

When analyzing the natural rubber with MCC, it was found that the distribution of particles was proportional to the amount of filler (Figure 2). When NCC was added to the rubber, there were some parts of the NCC that were evenly distributed and some agglomerated together according to the amount of filler. When more fillers were added, the agglomerate became larger. Therefore, there was a filler–filler interaction based on the specific surface area of nanocalcium carbonate, in particular, resulting in filler agglomeration of NCC (Figure 2).

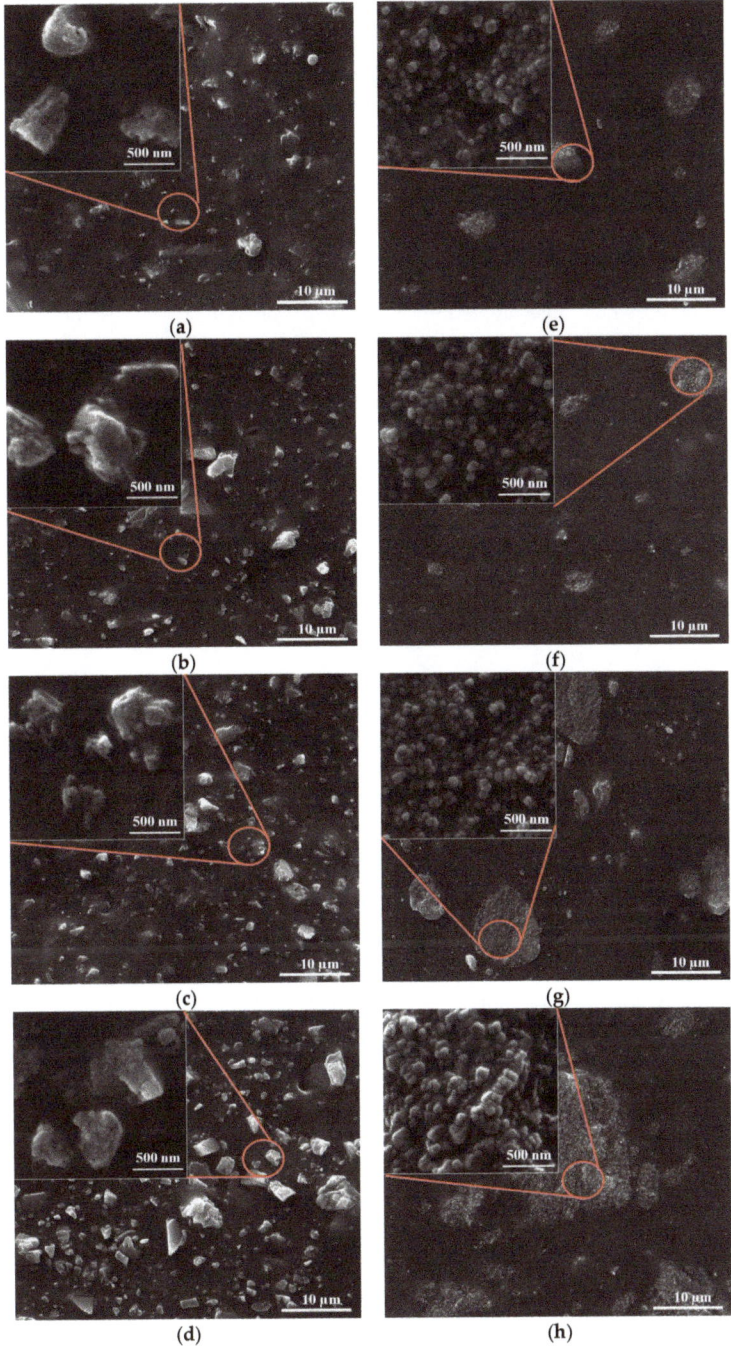

Figure 2. SEM images of rubber composites: (**a**) NR/MCC 25 phr, (**b**) NR/MCC 50 phr, (**c**) NR/MCC 75 phr, (**d**) NR/MCC 100 phr, (**e**) NR/NCC 25 phr, (**f**) NR/NCC 50 phr, (**g**) NR/NCC 75 phr, and (**h**) NR/NCC 100 phr.

3.2. Physical Properties

The Mooney viscosity test was performed in order to study the relationship between the macromolecular structure of NR filled with calcium carbonate and viscosity. Figure 3 shows the Mooney viscosity of natural rubber with either MCC or NCC. It was found that the Mooney viscosity of the rubber composites increased with increasing amount of either MCC or NCC into natural rubber. Because MCC and NCC are fillers that have solid particles, when mixed with natural rubber, the fillers can increase the Mooney viscosity of the rubber composite. When a filler (rigid material) is mixed into a rubber (soft material), the filler blocks the flow of rubber, which is called a "hydrodynamic effect" [27]. This causes the increase in Mooney viscosity of the rubber composites. Comparing MCC and NCC, the nanosized filler NCC had a much higher ratio of surface area per volume than the microsized filler MCC. This would probably cause higher movement restriction in the rubber composites. Therefore, more torques were required in the Mooney viscosity test of NCC blended rubber.

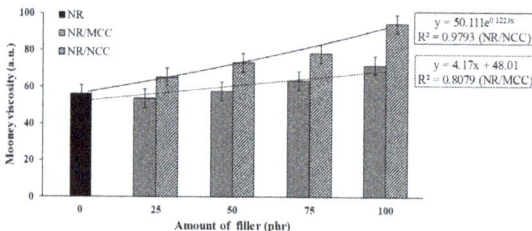

Figure 3. Mooney viscosity and linear relationship of natural rubber with either MCC or NCC at 25, 50, 75, and 100 phr.

The Mooney viscosity of the samples was almost stable when MCC at both 25 and 50 phr was added into NR. However, when the MCC was more than 50 phr, the Mooney viscosity of the rubber composites increased. In contrast, the Mooney viscosity of NR with NCC exponentially increased when the NCC was only at 25 phr, indicating that the smaller filler was able to interact with rubber more effectively (Figure 3). This result is in good agreement with a previous study, where the viscosity of epoxy composites supplemented with either graphene nanoplatelets or graphite was found to exponentially increase with the loading of filler [28].

We studied the interaction of rubber–filler using bound rubber, which was measured by extraction of the rubber-free chain in the composite with toluene at room temperature [14]. The percentages of bound rubber in all samples are summarized in Figure 4. The bound rubber of NR/MCC linearly increased with increasing MCC amount, while the bound rubber of NR/NCC exponentially increased with increasing NCC amount, certainly due to the enhancement of NR and nanosized filler interaction. This result is in good agreement with the result of Mooney viscosity.

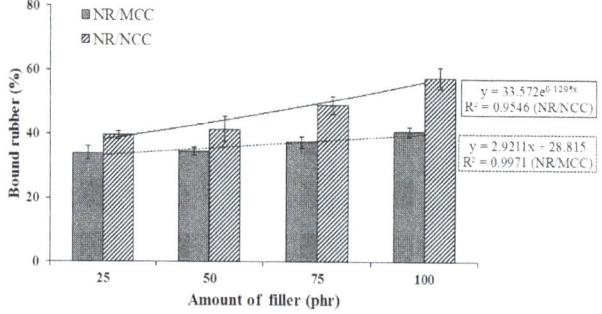

Figure 4. Bound rubber for natural rubber (NR) with various sizes (MCC vs. NCC) and amounts of filler (25, 50, 75, and 100 phr).

3.3. Mechanical Properties

Figure 5 demonstrates the stress–strain relationship from the experimental data and hyperelastic models of the NR system. For the reduced polynomial model used in this work, the 6th order is the best-fitting curve with experimental data. Table 2 shows the parameters from the fitting results of 6th order, and C_{10} is related to the initial shear modulus by $\mu_0 = 2C_{10}$, which are also shown in the same table. The relationships of the initial shear modulus of NR composites are illustrated in Figure 6. It can be seen that the initial shear modulus of NR/NCC was higher than that of NR/MCC at a given loading of filler. This result is in good agreement with the results of Mooney viscosity and bound rubber.

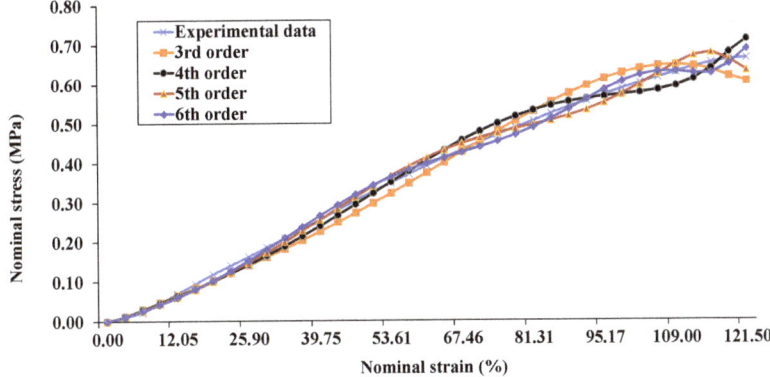

Figure 5. The stress–strain relationship from experiment data and hyperelastic modelling, 3rd to 6th order reduced polynomial.

Table 2. Calibrated parameters of 6th order of reduced polynomial for the rubber composites.

	C_{10} (10^{-3})	C_{20} (10^{-7})	C_{30} (10^{-10})	C_{40} (10^{-14})	C_{50} (10^{-18})	C_{60}	μ_0 (10^{-3})
NR	2.12	6.59	−1.60	1.79	−0.95	0.0	4.24
NR/MCC 25 phr	3.01	0.02	−0.39	0.55	−0.32	0.0	6.01
NR/MCC 50 phr	3.55	2.71	−1.14	1.41	−0.77	0.0	7.10
NR/MCC 75 phr	3.47	4.90	−1.71	2.08	−1.13	0.0	6.94
NR/MCC 100 phr	3.83	6.10	−2.07	2.50	−1.36	0.0	7.65
NR/NCC 25 phr	3.41	6.05	−1.96	2.33	−1.25	0.0	6.82
NR/NCC 50 phr	4.15	2.96	−1.40	1.80	−1.00	0.0	8.31
NR/NCC 75 phr	4.60	3.39	−1.40	1.82	−1.03	0.0	9.19
NR/NCC 100 phr	4.63	7.64	−2.53	3.07	−1.67	0.0	9.27

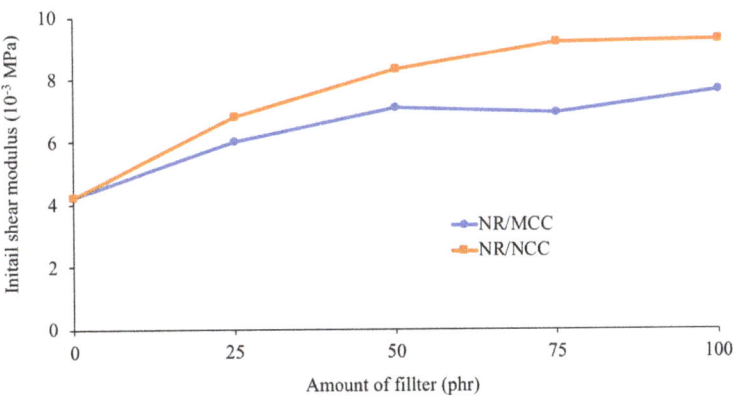

Figure 6. The initial shear modulus, μ_0, at different loading of MCC and NCC.

The Mullins effect is a phenomenon observed in rubber composite where the stress–strain equilibrium causes a strain-induced softening character between rubber and filler [29]. Forces are repeatedly applied to a rubber composite, which leads to weakening of the sample. In this study, the Mullins effect was analyzed by stress–strain curves (three loops of go-return curves) for natural rubber with addition of either MCC or NCC at 25, 50, 75, and 100 phr (Figure 7). We found that the Mullins effect of NR/MCC was stable below 50 phr of MCC; beyond this concentration, the Mullins effect of NR/MCC was more pronounced. The shape of Mullins effect of NR/NCC also changed above 50 phr of NCC. Meanwhile, NR/NCC had a higher Mullins effect than NR/MCC at a given concentration of filler, indicating that the high agglomeration of nanoparticle (NCC) contributes to the high Mullins effect. This filler agglomeration can hinder the mobility of the macromolecular chains when pulling; thus, higher stress concentration at the localized spot can occur [30]. The microsized $CaCO_3$ could easily slip between macromolecular chains in the NR matrix, so it could decrease the stress softness. For all the samples, the stress–strain curves of the second and third loops were always less pronounced in Mullins effect compared to the first loop.

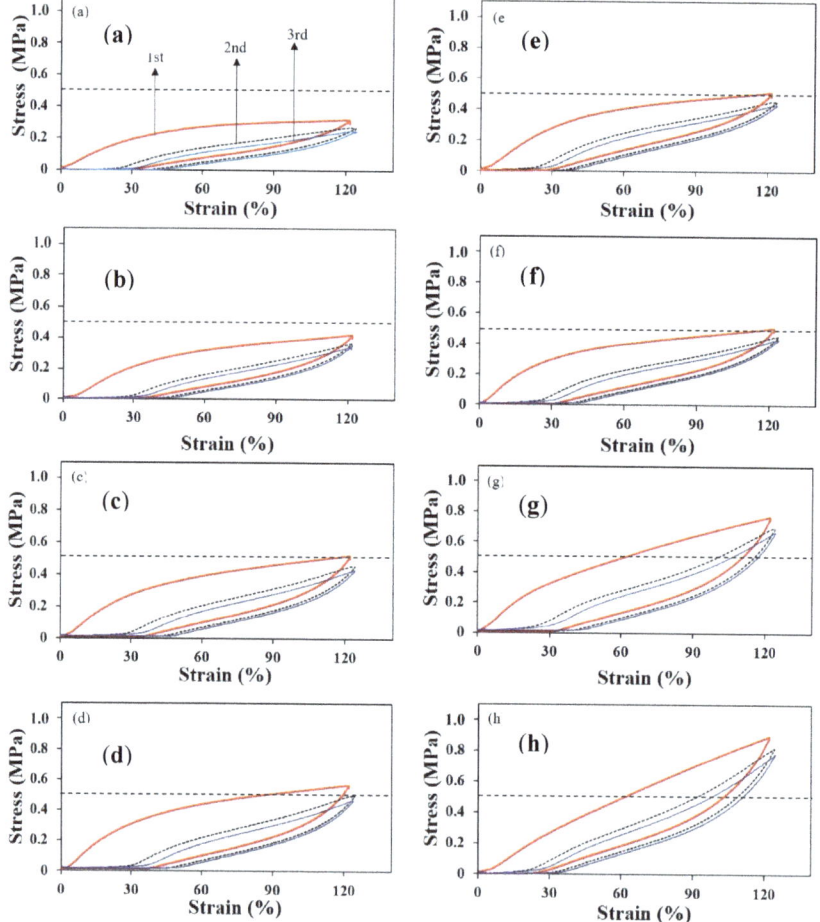

Figure 7. Mullins effect of NR/MCC (**left**) and NR/NCC (**right**) composites when the amount of either MCC or NCC were at (**a,e**) 25 phr, (**b,f**) 50 phr, (**c,g**) 75 phr, and (**d,h**) 100 phr.

Based on the results of the Mullins effect, we were also interested to investigate the stress relaxation of rubber molecules filled with either MCC or NCC, which can occur through viscoelastic flow or slippage of entanglements loosening the network of rubber chains or may arise from scission of the rubber chains supporting the stress. Stress relaxation of a composite occurs when a constant stress creates physical and/or chemical changes between the rubber molecule and the filler, thus reducing the force that the composite exerts over a certain period of time. Then, the relaxation of the rubber chains and fillers can be presented. If the changes that occur are generally chemical reactions, the effects tend to be long-term and irreversible [31,32].

Figure 8a shows the stress relaxation curves of pure NR and NR/MCC composites, and Figure 8b shows the stress relaxation curves of pure NR and NR/NCC composites. We plotted a fitted curve and then extracted the values in order to form a negative exponential equation $Y = Ce^{-kX} + Y_o$, where C is the difference value between initial stress and final stress, k is the rate of stress relaxation, and Yo is the final stress at time 0. We found that the rates of stress relaxation (k) for both types of filler (MCC and NCC) increased with increasing filler loading. This results is in good agreement with a previous study of organically modified montmorillonite-filled natural rubber/nitrile rubber nanocomposites, where the rate of stress relaxation was also found to increase with increasing filler loading [33] due to the increase in interaction, thus being more pronounced in entropy.

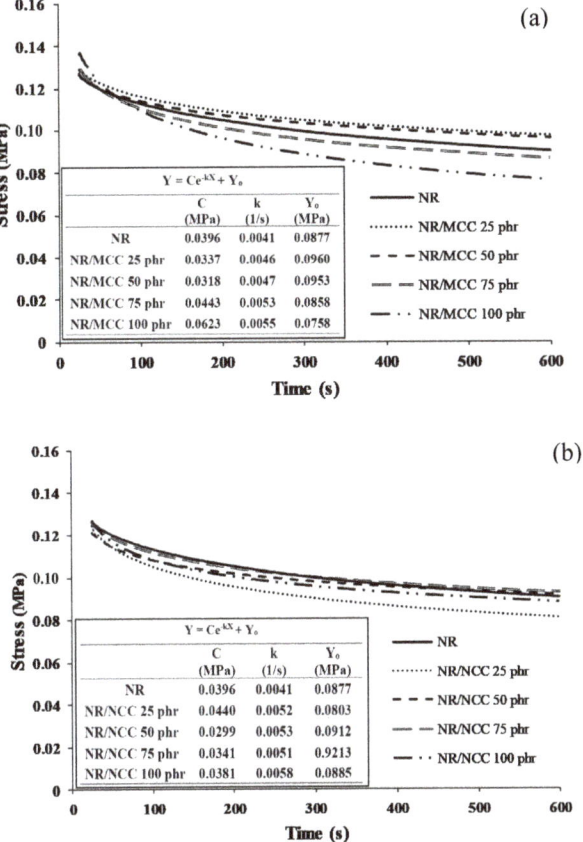

Figure 8. Stress relaxation curves of samples: (a) pure NR and NR with MCC at 0, 25, 50, 75, and 100 phr, (b) pure NR and NR with NCC at 0, 25, 50, 75, and 100 phr.

Focusing on the filler–filler interaction and particularly the effect of calcium carbonate size on composite properties, next, we examined properties of the composites in shear mode of the rubber processing analyzer. The Payne effect is a feature of the stress–strain behavior, which is related to the shape change caused by strain in the rubber with fillers [34]. This phenomenon is related to the storage modulus (G′) and loss modulus (G″) in shear deformation conditions. The reason for this phenomenon is the formation of a network formed by filler–filler interaction in the composites at low strain. In real life, this means the energy loss (tan delta) and the efforts are to minimize this Payne effect in filled rubber.

Figure 9 shows the Payne effect of NR/MCC composites. We found that NR/MCC at 25 phr of MCC possessed the lowest value of different storage moduli (ΔG′), certainly due to the weaker filler network, which resulted in the largest interaggregate distance in the rubber matrix. Meanwhile, NR/MCC at 100 phr of MCC possessed the highest value of different storage moduli (ΔG′). This Payne effect can be applied to explain the destruction–reformation of filler–filler networks and adsorption–desorption of rubber chains at the filler interface of the rubber composite [21]. The fact is that upon loading of MCC, the interaggregate distances become smaller with increasing filler content, and the probability of the formation of a filler network therefore increases. The damping peak (tan delta) of the composites with MCC continuously increased with increasing strain, with higher value of ΔG′ possessing higher damping peak or hysteresis for the same sample system (Figure 9). Therefore, the hysteresis resulted from the breakdown of the filler network, and the straining disruption could dissipate energy. Concerning the NR/NCC composites (Figure 10), their values of ΔG′ were higher than those of NR/MCC composites. Moreover, the values of ΔG′ for NR/NCC composites exponentially increased with increasing filler content, whereas the NR/MCC composites possessed a linear increase of ΔG′ with increasing filler content. This can be explained by the pronounced effect of the filler–filler network of NCC. Interestingly, NR/NCC composites had more filler–filler interaction or agglomeration than NR/MCC composites at a given loading of filler, while the tan delta of NR/NCC composites was lower than that of NR/MCC composites. This might have come from the synergy effect between bound rubber and the Payne effect. NR/NCC composites had higher bound rubber or network of rubber–filler interaction than NR/MCC composites, so the NR/NCC composites possessed lower heat build-up within the material. Higher rubber networks possess lower dissipation of energy or tan delta [2].

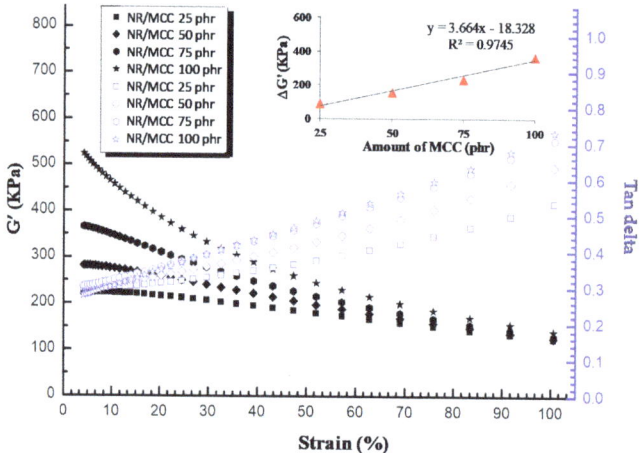

Figure 9. Payne effect (**left**; solid) and damping factor (tan delta, **right**; open) as a function of strain for NR with MCC particles at 25–100 phr.

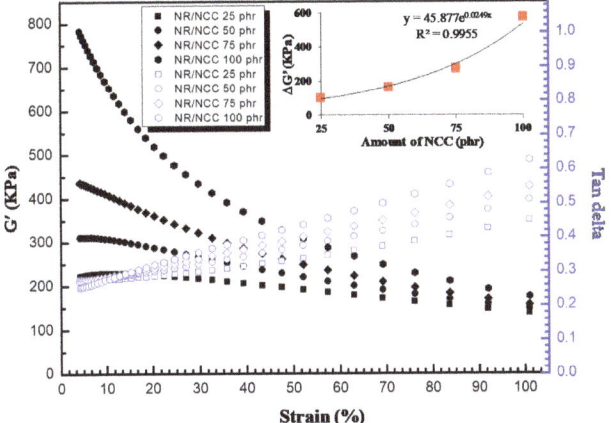

Figure 10. Payne effect (**left**; solid) and damping factor (tan delta, **right**; open) as a function of strain for NR with NCC particles at 25–100 phr.

Then, we also estimated the reinforcement efficiency of the filler in NR matrix (α_f) as a modified equation below [35]:

$$\alpha_f = \frac{\frac{M_f}{M_g} - 1}{w} \tag{6}$$

where M_f and M_g are the Mooney viscosity for the filled composite and pure gum, respectively, and W is the mass fraction of the filler in the composites. This parameter (α_f) is a measurement of the filler activity in the polymer matrix, so higher numerical value of α_f indicates higher polymer–filler interactions [34].

We found that the reinforcement efficiency of NR/NCC composites was higher than that of NR/MCC composites (Figure 11). This result is in good agreement with the result of the Payne effect. There was no reinforcement efficiency of NR/MCC at 25 phr. However, we applied this equation for the Mooney viscosity of green rubber composites compared to the original equation, which uses the torque from rheometer of crosslinked composites [35].

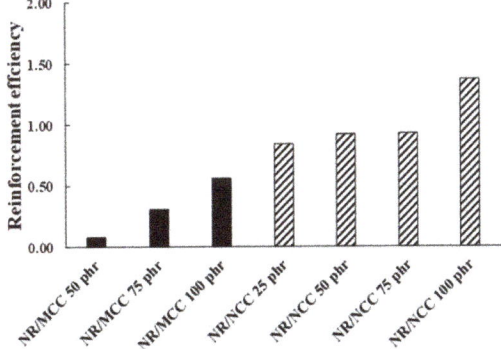

Figure 11. Filler reinforcing efficiency in different NR composites.

Master curves (time–temperature superposition) was determined by dynamic mechanical analysis, which represents the storage modulus of rubber samples as a function of reduced frequency. Table 3 shows the mean values of the constants C_1 and C_2 for universal, Ferry, filled composites and pure rubber, respectively. These two constant values are in good agreement with the Ferry reference and previous studies [23,24,36]. All the composite samples possessed the same C_1, C_2, and shift factor (a_T) values as those of the pure NR sample [36].

Table 3. The constants C_1 and C_2 for calculating the shift factor [23,24].

Reference	C_1	C_2
Universal	17.44	51.60
Ferry	5.94	151.60
All NR samples	8.50	186.50

Both Figure 12 (NR/MCC) and Figure 13 (NR/NCC) present the master curves of rubber composites in three zones: glassy plateau at high reduced frequency, transition state, and rubbery plateau at low reduced frequency. It can be seen that all the rubbers had almost similar viscoelastic properties irrespective of whether NR was filled with MCC or NCC at different temperatures and frequencies. The viscoelastic properties of NR composites at low temperature was equivalent to those of NR composites at high frequency and vice versa. However, we also found that NR/MCC and NR/NCC at glassy state were influenced by the addition of large quantity of fillers, indicating the lower free volume in the composites for higher loading of filler. For the rubbery state, composites with higher loading of filler had higher level in this rubbery plateau, indicating more relaxation time of rubber molecules.

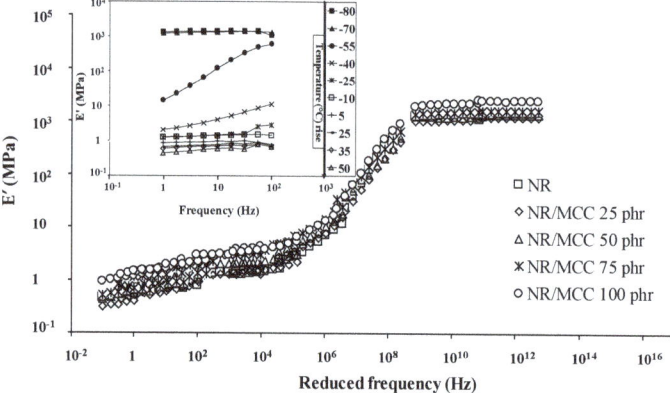

Figure 12. Elastic modulus (E') versus reduced frequency master curves for pure NR and NR/MCC composites. Insert figure presents raw data of pure NR before establishing the master curve.

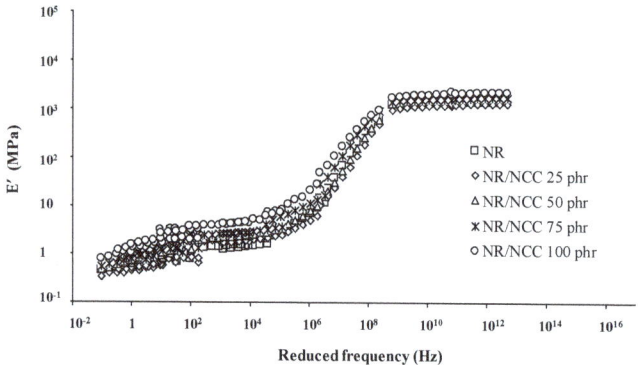

Figure 13. Elastic modulus (E') versus reduced frequency master curves for pure NR and NR/NCC composites.

4. Conclusions

This work focused on investigating the influence of calcium carbonate morphological parameters (polymorph and different particle sizes) on the properties of uncrosslinked rubber composite, in particular, new results of structure–properties relationship. The different size of calcium carbonate has an influence on the static and dynamic mechanical properties of rubber composites corresponding to the polymorph and its particle sizes. Hence, based on the results of this study, the SEM images of MCC and NCC particles showed the same spherical shape with different sizes. Finite element analysis of the rubber composites was in good agreement with the results of experimental data. The Mullins effect of rubber composites filled with either MCC or NCC was in good agreement with the results of Mooney viscosity and bound rubber, with higher Mooney viscosity and bound rubber leading to higher stress to pull the rubber composites. The rate of stress relaxation increased with increasing filler loading for both MCC and NCC. The Payne effect showed that the value of different storage moduli ($\Delta G'$) of rubber composites filled with 25 phr NCC was the lowest due to weaker filler network, while the NR supplemented with 100 phr NCC had more significant $\Delta G'$ with the increase in strain. The results of rubber composites filled with MCC showed the same tendency as those of rubber composites filled with NCC. However, the effect of specific surface area of NCC on the properties of rubber composites was more pronounced than those of rubber composites filled with MCC. The master curves from WLF superposition showed similar viscoelastic properties for both NR/MCC and NR/NCC composites. This research elucidates that we can utilize the optimum amount of nanocalcium carbonate (25–100 phr depending on the product's requirement) in order to enhance the properties and processability of natural rubber composites.

Author Contributions: N.P. carried out the preparation and characterization of rubber composites; W.P. carried out the finite element analysis; S.P. contributed to the discussion of the manuscript; W.S. conceived the study, designed the study, and worked on the manuscript. All authors gave final approval for publication and agree to be held accountable for the work performed herein. All authors have read and agreed to the published version of the manuscript.

Funding: This research received no external funding.

Acknowledgments: We are thankful to Thai Synthetic Rubbers Co., Ltd. for the test of Payne effect and Sand and Soil Industry Co., Ltd. for nano- and microcalcium carbonate support. This work was partly supported by Siam United Rubber Co., Ltd.

Conflicts of Interest: The authors declare no conflict of interest in reported research. The funders had no role in the design of the study; in the collection, analyses, or interpretation of data; in the writing of the manuscript; or in the decision to publish the results.

References

1. Smitthipong, W.; Suethao, S.; Shah, D.; Vollrath, F. Interesting green elastomeric composites: Silk textile reinforced natural rubber. *Polym. Test.* **2016**, *55*, 17–24.
2. Chollakup, R.; Suwanruji, P.; Tantatherdtam, R.; Smitthipong, W. New approach on structure-property relationships of stabilized natural rubbers. *J. Polym. Res.* **2019**, *26*, 37. [CrossRef]
3. Suksup, R.; Sun, Y.; Sukatta, U.; Smitthipong, W. Foam rubber from centrifuged and creamed latex. *J. Polym. Eng.* **2019**, *39*, 336–342. [CrossRef]
4. Kohjiya, S.; Ikeda, Y. Reinforcement of general-purpose grade rubbers by silica generated in situ. *Rubber Chem. Technol.* **2000**, *73*, 534–550. [CrossRef]
5. Findik, F.; Yilmaz, R.; Köksal, T. Investigation of mechanical and physical properties of several industrial rubbers. *Mater. Des.* **2004**, *25*, 269–276. [CrossRef]
6. Granda, L.A.; Oliver-Ortega, H.; Fabra, M.J.; Tarres, Q.; Pelach, M.A.; Lagaron, J.M.; Mendez, J.A. Improved process to obtain nanofibrillated cellulose (CNF) reinforced starch films with upgraded mechanical properties and barrier character. *Polymers* **2020**, *12*, 1071. [CrossRef]
7. Zare, Y.; Rhee, K.Y. Model progress for tensile power of polymer nanocomposites reinforced with carbon nanotubes by percolating interphase zone and network aspects. *Polymers* **2020**, *12*, 1047. [CrossRef]

8. Prioglio, G.; Agnelli, S.; Conzatti, L.; Balasooriya, W.; Schrittesser, B.; Galimberti, M. Graphene layers functionalized with a Janus pyrrole-based compound in natural rubber nanocomposites with improved ultimate and fracture properties. *Polymers* **2020**, *12*, 944. [CrossRef]
9. Jin, J.; Noordermeer, J.W.M.; Dierkes, W.K.; Blume, A. The effect of silanization temperature and time on the marching modulus of silica-filled tire tread compounds. *Polymers* **2020**, *12*, 209. [CrossRef]
10. Roberson, C.G.; Tunnicliffe, L.B.; Maciag, L.; Bauman, M.A.; Miller, K.; Herd, C.R.; Mars, W.V. Characterizing distributions of tensile strength and crack precursor size to evaluate filler dispersion effects and reliability of rubber. *Polymers* **2020**, *12*, 203. [CrossRef]
11. Abate, L.; Bottino, F.A.; Cicala, G.; Chiacchio, M.A.; Ognibene, G.; Blamco, I. Polystyrene nanocomposites reinforced with novel dumbbell-shaped phenyl-POSSs: Synthesis and thermal characterization. *Polymers* **2019**, *11*, 1475. [CrossRef] [PubMed]
12. Hejna, A.; Barczewski, M.; Andrzejewski, J.; Kosmela, P.; Piasecki, A.; Szostak, M.; Kuang, T. Rotational molding of linear low-density polyethylene composites filled with wheat bran. *Polymers* **2020**, *12*, 1004. [CrossRef] [PubMed]
13. Toth, L.F.; Baets, P.; Szebenyl, G. Thermal, viscoelastic, mechanical and wear behavior of nanoparticle filled polytetrafluoroethylene: A comparison. *Polymers* **2020**, *12*, 1940. [CrossRef] [PubMed]
14. Khan, I.; Bhat, A.H. Micro and nano calcium carbonate filled natural rubber composites and nanocomposites. *Nat. Rubber Mater.* **2013**, *2*, 467–487.
15. Prasertsri, S.; Vudjung, C.; Inthisaeng, W.; Srichan, S.; Sapprasert, K.; Kongon, J. Comparison of reinforcing efficiency between calcium carbonate/carbon black and calcium carbonate/silica hybrid filled natural rubber composites. *Defect Diffus. Forum* **2018**, *382*, 94–98. [CrossRef]
16. Gopinath, C.S.; Hegde, S.G.; Ramaswamy, A.V.; Mahapatra, S. Photoemission studies of polymorphic $CaCO_3$ materials. *Mater. Res. Bull.* **2002**, *37*, 1323–1332. [CrossRef]
17. Meng, L.; Wang, J.; Liu, Q.; Fan, Z. Hydrophobic calcium carbonate with hierarchical micro-/nanostructure for improving foaming capacity. *Mater. Res. Express* **2020**, *6*, 1250–1258. [CrossRef]
18. Wang, C.; Zhao, J.; Zhao, X.; Bala, H.; Wang, Z. Synthesis of nanosized calcium carbonate (aragonite) via a polyacrylamide inducing process. *Powder Technol.* **2006**, *163*, 134–138. [CrossRef]
19. Wu, D.; Wang, X.; Song, Y.; Jin, R. Nanocomposites of poly(vinyl chloride) and nanometric calcium carbonate particles: Effects of chlorinated polyethylene on mechanical properties, morphology, and rheology. *J. Appl. Polym. Sci.* **2004**, *92*, 2714–2723. [CrossRef]
20. Fang, Q.; Song, B.; Tee, T.-T.; Sin, L.; Hui, D.; Bee, S.-T. Investigation of dynamic characteristics of nano-size calcium carbonate added in natural rubber vulcanizate. *Compos. Part B Eng.* **2014**, *60*, 561–567. [CrossRef]
21. Liu, W.; Yang, Z.; Zheng, Y.; Wang, H. The effect of OMNT on the properties of vehicle damping carbon black-natural rubber composites. *Polymers* **2020**, *12*, 1983. [CrossRef]
22. Yan, X.; Hamed, G.R.; Jia, L. Modulating silica-rubber interface by a biorenewable urushiol derivative. Synthesis, surface modification, and mechanical and dynamic mechanical properties of vulcanizates therefrom. *J. Appl. Polym. Sci.* **2018**, *135*, 45937. [CrossRef]
23. Smitthipong, W.; Nardin, M.; Schultz, J.; Nipithakul, T.; Suchiva, K. Study of tack properties of uncrosslinked natural rubber. *J. Adhes. Sci. Technol.* **2004**, *18*, 1449–1463. [CrossRef]
24. Williams, M.L.; Landel, R.F.; Ferry, J.D. The temperature dependence of relaxation mechanisms in amorphous polymers and other glass-forming liquids. *J. Am. Chem. Soc.* **1955**, *77*, 3701–3707. [CrossRef]
25. Mase, G.; Ronald, E.; George, E. *Continuum Mechanics for Engineers*; CRC Press: Boca Raton, FL, USA, 2009.
26. Wedin, P.; Lewis, J.A.; Bergstrom, L. Soluble organic additive effects on stress development during drying of calcium carbonate suspension. *J. Colloid Interface Sci.* **2005**, *290*, 134–144. [CrossRef]
27. Fröhlich, J.; Niedermeier, W.; Luginsland, H.D. The effect of filler–filler and filler–elastomer interaction on rubber reinforcement. *Compos. Part A Appl. Sci. Manuf.* **2005**, *36*, 449–460. [CrossRef]
28. Levy, I.; Wormser, E.M.; Varenik, M.; Buzaglo, M.; Nadiv, R.; Regev, O. Graphene–graphite hybrid epoxy composites with controllable workability for thermal management. *Beilstein J. Nanotechnol.* **2019**, *10*, 95–104. [CrossRef]
29. Mullins, L. Effect of Stretching on the Properties of Rubber. *Rubber Chem. Technol.* **1948**, *21*, 281–300. [CrossRef]

30. Bee, S.T.; Hassan, A.; Ratnam, C.T.; Tee, T.T.; Sin, L.T. Effects of montmorillonite on the electron beam irradiated alumina trihydrate added polyethylene and ethylene vinyl acetate nanocomposite. *Polym. Compos.* **2012**, *33*, 1883–1892. [CrossRef]
31. Robinson, H.W.H.; Vodden, H.A. Stress relaxation in rubber-evaluation of antioxidants. *Ind. Eng. Chem.* **1955**, *47*, 1477–1481. [CrossRef]
32. Da Rocha, E.B.D.; Linhares, F.N.; Gabriel, C.F.S.; de Sousa, A.M.F.; Furtado, C.R.G. Stress relaxation of nitrile rubber composites filled with a hybrid metakaolin/carbon black filler under tensile and compressive forces. *Appl. Clay Sci.* **2018**, *151*, 181–188. [CrossRef]
33. Maria, H.J.; Lyczko, N.; Nzihou, A.; Joseph, K.; Mathew, C.; Thomas, S. Stress relaxation behavior of organically modified montmorillonite filled natural rubber/nitrile rubber nanocomposites. *Appl. Clay Sci.* **2014**, *87*, 120–128. [CrossRef]
34. Payne, A.R.; Whittaker, R.E. Low strain dynamic properties of filled rubbers. *Rubber Chem. Technol.* **1971**, *44*, 440–478. [CrossRef]
35. Ghari, H.S.; Jalali-Arani, A. Nanocomposites based on natural rubber, organoclay and nano-calcium carbonate: Study on the structure, cure behavior, static and dynamic-mechanical properties. *Appl. Clay Sci.* **2016**, *119*, 348–357. [CrossRef]
36. Promhuad, K.; Smitthipong, W. Effect of stabilizer states (solid vs. liquid) on properties of stabilized natural rubbers. *Polymers* **2020**, *12*, 741. [CrossRef]

© 2020 by the authors. Licensee MDPI, Basel, Switzerland. This article is an open access article distributed under the terms and conditions of the Creative Commons Attribution (CC BY) license (http://creativecommons.org/licenses/by/4.0/).

Article

Effect of Chlorophyll Hybrid Nanopigments from Broccoli Waste on Thermomechanical and Colour Behaviour of Polyester-Based Bionanocomposites

Bàrbara Micó-Vicent [1,2], Marina Ramos [3], Francesca Luzi [4], Franco Dominici [4], Valentín Viqueira [1], Luigi Torre [4], Alfonso Jiménez [3], Debora Puglia [4,*] and María Carmen Garrigós [3,*]

1. Colour and Vision Group, University of Alicante, San Vicente del Raspeig, ES-03690 Alicante, Spain; barbara.mico@ua.es (B.M.-V.); valentin.viqueira@ua.es (V.V.)
2. Department of Appl. Stat. & Operat. Research, & Qual., Universitat Politècnica de València, ES-03801 Valencia, Spain
3. Department of Analytical Chemistry, Nutrition & Food Sciences, University of Alicante, San Vicente del Raspeig, ES-03690 Alicante, Spain; marina.ramos@ua.es (M.R.); alfjimenez@ua.es (A.J.)
4. Department of Civil and Environmental Engineering, University of Perugia, 05100 Terni, Italy; francesca.luzi@unipg.it (F.L.); francodominici1@gmail.com (F.D.); luigi.torre@unipg.it (L.T.)
* Correspondence: debora.puglia@unipg.it (D.P.); mc.garrigos@ua.es (M.C.G.)

Received: 13 October 2020; Accepted: 26 October 2020; Published: 28 October 2020

Abstract: Natural dyes obtained from agro-food waste can be considered promising substitutes of synthetic dyes to be used in several applications. With this aim, in the present work, we studied the use of chlorophyll dye (CD) extracted from broccoli waste to obtain hybrid nanopigments based on calcined hydrotalcite (HT) and montmorillonite (MMT) nanoclays. The synthesized chlorophyll hybrid nanopigments (CDNPs), optimized by using statistical designed experiments, were melt-extruded with a polyester-based matrix (INZEA) at 7 wt% loading. Mechanical, thermal, structural, morphological and colour properties of the obtained bionanocomposites were evaluated. The obtained results evidenced that the maximum CD adsorption into HT was obtained when adding 5 wt% of surfactant (sodium dodecyl sulphate) without using any biomordant and coupling agent, while the optimal conditions for MMT were achieved without adding any of the studied modifiers. In both cases, an improvement in CD thermal stability was observed by its incorporation in the nanoclays, able to protect chlorophyll degradation. The addition of MMT to INZEA resulted in large ΔE^* values compared to HT incorporation, showing bionanocomposite green/yellow tones as a consequence of the CDNPs addition. The results obtained by XRD and TEM revealed a partially intercalated/exfoliated structure for INZEA-based bionanocomposites, due to the presence of an inorganic filler in the formulation of the commercial product, which was also confirmed by TGA analysis. CDNPs showed a reinforcement effect due to the presence of the hybrid nanopigments and up to 26% improvement in Young's modulus compared to neat INZEA. Finally, the incorporation of CDNPs induced a decrease in thermal stability as well as limited effect in the melting/crystallization behaviour of the INZEA matrix. The obtained results showed the potential use of green natural dyes from broccoli wastes, adsorbed into nanoclays, for the development of naturally coloured bionanocomposites.

Keywords: broccoli waste; chlorophyll; hybrid nanopigment; experimental design; nanoclays; bionanocomposites

1. Introduction

Natural dyes obtained from agro-food waste can be considered promising substitutes of synthetic dyes to be used in several applications [1], due to their easy availability, non-toxicity, antioxidant/antimicrobial activity, medicine applications and biodegradation capability showing no environmental issues [2–4]. In particular, broccoli wastes contain several nutrients and bioactive compounds such as pigments, glucosinolates and phenolic compounds, the valorisation of these wastes being very valuable for a wide range of industrial applications [5–7]. So, broccoli discards could be used as a potential source for the extraction of natural green dyes, including carotenoids (16–2689 µg g^{-1} DW) and chlorophylls (139–5569 µg g^{-1} DW), as already reported by Ferreira et al. [5]. However, natural dyes show some drawbacks such as challenging colour reproduction between different natural dye samples, low light fastness, restricted colour array, poor colour fastness and low chemical and thermal stability [8]. Chlorophyll dyes have been used in complex applications such as solar cell dyes due to their interesting properties [9,10], but the instability of these dyes under the exposure to oxygen, high temperature or light environments could limit their final applications [11,12].

Natural dyes are widely used in organic polymer formulations as colouring agents as well as to enhance other specific properties such as durability, mechanical properties and corrosion resistance. Different stabilization techniques such as microencapsulation and nanoclay adsorption are necessary to enhance the stability of natural dyes [13,14]. The use of modifiers, such as surfactants, silane agents and salts has been reported to improve nanoclay adsorption for different molecular dye structures [15–19]. Biomordant salts have been recently proposed as potential alternative additives to replace synthetic mordant salts in textile dyeing [20]. Hybrid dye-clay nanopigments have been reported to show good colour characteristics, excellent stability and mechanical performance [8,21,22]. Therefore, by combining the colour of natural dyes and the resistance of inorganic nanostructures, it is possible to obtain hybrid nanomaterials with improved chemical properties and higher stability [23]. The adsorption and intercalation mechanisms of reactive dyes within nanoclays for the development of nano-structured systems has been studied by different authors [24,25].

To the best of our knowledge, a minimal number of works have studied the synthesis of chlorophyll hybrid nanopigments [26] and their incorporation in polymer matrices. Specifically, while some authors investigated the role of these nanopigments in photodynamic therapy [27] and energy transfer in dye-sensitized solar cells [28], few authors investigated the effect of their introduction in polymer matrices. Ahmed et al. [29] showed that chlorophyll and anthocyanin could act as plasticizers affecting secondary bonds of polyvinyl chloride (PVC) polymer. In another work, Chandrappa et al. [30] successfully incorporated green tricolour leaf extract into a polyvinyl alcohol (PVA) matrix to obtain PVA-based biofunctional composite structures, showing the potential use of these biocomposites in UV shielding protective sheets/layers, windows or coatings for terrestrial and aquatic ecosystems. The smart response of chlorophyll, when combined with polypyrrole and gluten, was also proved by other authors, obtaining wheat gluten/chlorophyll/polypyrrole nanocomposite smart films due to their conductivity and colour change; additionally, chlorophyll pigment can enhance gluten film properties showing high potential for improving food shelf-life of packaged food [31,32]. In all these cases, dyes were introduced in their free state without using the hybrid combination with inorganic nanostructures.

The incorporation of hybrid nanopigments into polymer matrices could improve the physical and mechanical properties of the final nanocomposites [8,33,34]. Micó-Vicent et al. [19,22] demonstrated that the incorporation of hybrid nanopigments based on chlorophyll, beta-carotene and beetroot extract natural dyes into an epoxy bioresin improved the optical performance, thermal stability and UV-Vis light exposure stability of the natural dyes and bioresins. Mahmoodi et al. [35] reported the synthesis of hybrid dye-clay nanopigments and its application in epoxy coatings observing that the dyes chemically attached to the nanoclay particles do not migrate from the polymer matrix, and the resulting hybrid nanopigments showed better photo-stability and dispersion than that of pure dyes. The interesting properties of hybrid dye-clay nanopigments have been also applied in 3D printing technologies for the development of innovative coloured functional materials [36–38]. Mahmoodi et al. [39] also obtained coloured biodegradable/biocompatible poly(lactic acid) (PLA) nanocomposite films containing 1–5 wt% of a dye-clay hybrid nanopigment using a simple solution casting approach. High thermomechanical resistance as well as superior light and mass transport barriers for food packaging applications were observed for the developed nanocomposite films. Similar results were also obtained by Usopt et al. [40], who considered a dye containing chlorophyll from *Cassia alata* leaves as a colourant in blended poly(methylmetacrylate) (PMMA)-acrylic polyol for coating applications in the presence of copper (II) nitrate ($Cu(NO_3)_2$). The results from colour measurement showed that yellowness, glossiness and reflectivity were influenced by the natural dye, providing good colour stability in the produced films.

In this work, we propose the novel approach of synthesising chlorophyll hybrid nanopigments by using broccoli waste as a natural source and two nanoclay types with different ion exchange capacities. For this purpose, statistical experimental design experiments were considered to optimize the hybrid nanopigments preparation in the presence of different modifiers (surfactant, silane and natural biomordant obtained from pomegranate waste). We also verified, for the first time, how the incorporation of the produced chlorophyll hybrid nanopigments could enhance the overall behaviour (thermal, mechanical, morphological, structural and colour properties) of the bionanocomposites obtained by mixing the hybrid materials with a commercial biopolyester matrix.

2. Materials and Methods

2.1. Materials and Reagents

Broccoli waste, including leaves, stems and flowering parts, was obtained from discarded vegetables from FECOAM (Murcia, Spain). Broccoli waste was slightly cut into small pieces using a household blade cutter for 10 s at medium speed to reach particles of 1–50 mm^3 [8].

Montmorillonite (MMT, Gel White) and hydrotalcite (HT, BioUltra, ≥99.0%) laminar nanoclays, with a different charge ion capacity, were supplied by Southern Clay Products (Gonzales, TX, USA) and Sigma-Aldrich (St. Louis, MO, USA), respectively. HT was calcined at 600 °C for 3 h before use. Cetylpyridinium bromide (CPB) and sodium dodecyl sulphate (SDS) were used as surfactants for MMT and HT, respectively. A coupling agent (3-Aminopropyl) triethoxysilane (as silane) and a natural biomordant obtained from pomegranate waste peel (FECOAM, Murcia, Spain) were used as surface modifiers. All reagents and chemicals were of analytical grade, and they were purchased from Sigma-Aldrich (St. Louis, MO, USA).

INZEAF2 biopolyester commercial grade (density of 1.23 g cm^{-3} at 23 °C; moisture content < 0.5 %; melt flow rate of 19 g 10 min^{-1} (2.16 kg, 190 °C) was kindly provided by Nurel (Zaragoza, Spain) and used for bionanocomposites preparation.

2.2. Broccoli Dye and Pomegranate Peels Biomordant Extraction

A FLEXIWAVE™ microwave oven (Milestone srl, Bergamo, Italy) was used to obtain the natural additives used in this work, chlorophyll dye (CD) from broccoli wastes and pomegranate biomordant, by following methods previously optimized.

For CD, 20 g of freshly cut broccoli waste and 90% (v/v) acetone were used for microwave-assisted extraction (MAE) at a liquid to solid ratio of 5:1 in a 250 mL round-bottom flask. The sample was heated at 6 W g^{-1} for 25 min under reflux and stirring (400 rpm). The boiling temperature was reached at around 60 °C and maintained during extraction. After MAE, 5 mL of water containing 5 wt% of sodium carbonate (Na_2CO_3) was used to maintain the stability of the extracted chlorophylls.

To obtain the biomordant from pomegranate waste, peels were dried at 40 °C for 24 h in a climatic chamber (Dycometal, Barcelona, Spain) at a relative humidity (RH) of 25%. Dried peels were grounded with a high-speed rotor mill at 12,000 rpm (Ultra Centrifugal Mill ZM 200, RETSCH, Haan, Germany). Particles passing through a 0.5 mm sieve were used. MAE was performed by using 6 g of pomegranate sample mixed with 60 mL of 40% (v/v) ethanol (solid to solvent ratio of 1:10) for 10 min at 65 °C.

The obtained CD and biomordant extracts were filtered, and polysaccharide compounds were precipitated by adding 96% (v/v) ethanol. Samples were kept overnight at −20 °C to promote precipitation of insoluble compounds, and they were vacuum-filtered. The solvent present in the samples was firstly removed in a rotary evaporator (R-300, Büchi Labortechnik AG, Switzerland) and the aqueous solution was freeze-dried (LyoQuest Plus, Telstar, Terrassa, Spain). The purified additives were stored in the darkness at room temperature until further use.

2.3. Synthesis of Chlorophyll Hybrid Nanopigments (CDNPs)

The synthesis of the chlorophyll hybrid nanopigments (CDNPs) was performed by following the water/organic solvent dispersion method, based on previous studies [19]. A 2^{4-1} fractional factorial design of experiments (DoE) consisting of 8 experiments was used to study the best synthesis conditions to maximize broccoli dye adsorption into the laminar nanoclays. The solvent used in all experiments was 50 % (v/v) ethanol due to the low solubility of chlorophyll dye shown in water. The total amount of dye loaded for each sample was 0.2 g in 3 g of nanoclay. Four independent variables at two levels were considered (Table 1): nanoclay ion exchange capacity (HT or MMT), surfactant concentration (0 and 5 wt%), biomordant concentration (0 and 1 wt%) and silane concentration (0 and 5 wt%). The amount of intercalated dye in the nanoclay system was determined using a UV-Vis spectrophotometer (JASCO V650, Easton, MD, USA) at 405 nm by calculating the dye concentration present in the separated supernatants. Calibration curves were prepared in 50% (v/v) ethanol. The dye adsorbed over the initially added dye (%) was used as response to be maximized in the DoE analysis.

Table 1. 2^{4-1} fractional experimental design matrix and chlorophyll dye adsorbed over the initially added dye (%).

Experiment	Nanoclay	Surfactant (wt%) *	Biomordant (wt%)	Silane (wt%)	Adsorbed Dye (Ads, %)
1	HT	5	0	5	98.13
2	HT	0	0	0	98.44
3	HT	5	1	0	97.97
4	HT	0	1	5	98.12
5	MMT	5	0	0	97.83
6	MMT	0	0	5	97.52
7	MMT	0	1	0	95.66
8	MMT	5	1	5	96.62

* Cetylpyridinium bromide (CPB) was used for montmorillonite (MMT) and sodium dodecyl sulfate (SDS) for hydrotalcite (HT).

In addition, the final visual appearance of the optimal CDNPs and their thermal properties were evaluated using thermogravimetric analysis (see conditions in Section 2.5).

2.4. Bionanocomposites Preparation

Bionanocomposites based on INZEAF2 were obtained by melt blending the biopolymer and the synthetized CDNPs at 7 wt%. A co-rotating twin-screw extruder, Xplore 5 and 15 Micro Compounder by DSM, was used by mixing at a rotating speed of 90 rpm for 3 min, setting a temperature profile of 190–195–200 °C in the three heating zones from the feeding section to die. A Micro Injection Moulding Machine 10cc by DSM, coupled to the extruder and equipped with adequate moulds, was used to produce samples for tensile tests according to the standards. An appropriate pressure/time profile was used for the injection of each type of samples, while the temperatures of the injection barrel and the moulds were set, respectively, at 210 and 30 °C.

2.5. Bionanocomposite Characterization

Thermal degradation of CD and CDNPs as well as bionanocomposites was evaluated by thermogravimetric analysis (TGA, Seiko Exstar 6300, Tokyo, Japan). Around 5 mg of samples was heated from 30 to 800 °C at 10 °C min^{-1} under nitrogen atmosphere (200 mL min^{-1}). Three replicates of each sample were performed.

Differential scanning calorimetry (DSC) tests for bionanocomposites were conducted, in triplicate, using a DSC Q200 (TA Instruments, New castle, DE, USA) under nitrogen atmosphere (50 mL min^{-1}). Samples of 3 mg were introduced in aluminum pans (40 µL), and they were submitted to the following thermal program: −30 to 250 °C at 10 °C min^{-1}, with two heating scans and one cooling scan.

Dynamic mechanical thermal analysis (DMTA) of produced materials was performed, in triplicate, by using an Ares N2 rheometer (Rheometric Scientific, Epsom, Surrey, UK). Samples with dimensions of 2 × 5 × 40 mm, gripped with a gap of 20 mm, were tested in rectangular torsion at a frequency of 2π rad s^{-1} with a strain of 0.05%. A temperature ramp of 3 °C min^{-1} was applied in the range from 30 to 110 °C.

Optical properties of bionanocomposites were studied with a Konica Minolta sphere integrated spectrophotometer (CM-2300d, Tokyo, Japan). Data were acquired by using the SCI 10/D65 method, whereas CIELAB colour variables were used. Samples were placed on a white standard plate and L*, a* and b* parameters were determined. Measurements were performed, in triplicate, at random locations on each sample. Total colour difference $\Delta E_{ab}*$ was calculated with the obtained colorimetric attributes of the CIELAB colour space.

Tensile tests were carried out for bionanocomposites by using a universal test machine LK30 (Lloyd Instruments Ltd.) at room temperature according to ASTM D638-14 Standard. A minimum of five different samples was tested using a 5 kN load cell, setting the crosshead speed to 5 mm min^{-1}.

The CDNPs dispersion in the bionanocomposites was studied by transmission electron microscopy (TEM) and wide angle X-ray scattering (WAXS). TEM micrographs were obtained with a JEOL JEM-2010 (Tokyo, Japan) with an accelerating voltage of 100 kV. WAXS patterns were recorded at room temperature using a Bruker D8-Advance diffractometer (Madison, WI, USA) at scattering angles (2θ) 2.5°–80° (scanning rate of 3 s step^{-1} and step size of 5°) using filtered Cu Kα radiation (1.54 Å).

2.6. Statistical Analysis

Statistical analysis of results was performed by using Statgraphics Centurion XVI (Statistical Graphics, Rockville, MD, USA) to generate and analyse the results of the experimental design. The graphic analysis of the main effects and interactions between variables was used and the analysis of variance (ANOVA) was carried out. Differences between average values were assessed based on the Tukey test at a confidence level of 95% ($p < 0.05$). Characterization results for bionanocomposites were expressed as the mean ± standard deviation.

3. Results

3.1. Chlorophyll Hybrid Nanopigments (CDNPs)

The synthesis conditions for CDNPs were optimized by calculating the dye adsorption (Ads, %) as the concentration difference between the initially added dye in the nanoclay dispersion and the dye separated from the solvent after centrifugation. As can be seen in Table 1, good values for dye adsorption were obtained ranging from 95.66 to 98.44%, indicating a good degree of incorporation of the chlorophyll dye (CD) into HT and MMT nanoclays.

Analysis of variance (ANOVA) was performed to evaluate the effect of the studied variables on the dye adsorption (Table 2). A high degree of correlation between experimental and predicted values was obtained (R^2 = 99.98%). The statistical analysis showed that the synthesis of CDNPs was mainly influenced by the biomordant and silane modifiers (with a negative effect), followed by the surfactant which a positive effect—as can be seen in the Pareto and main effect plots (Figure 1).

Table 2. Variance analysis for the synthesis of chlorophyll hybrid nanopigments.

Factor	Sum of Squares	DF	Mean Square	F-Value	p-Value
B (Surfactant)	0.2387	1	0.2387	216.15	0.0432
C (Mordant)	2.8227	1	2.8227	2555.63	0.0126
D (Silane)	1.8126	1	1.8126	1641.11	0.0157
AB (Nanoclay-Surfactant)	0.8489	1	0.8489	768.59	0.0230
AC (Nanoclay-Mordant)	0.1624	1	0.1624	147.08	0.0524
AD (Nanoclay-Silane)	0.2549	1	0.2549	230.78	0.0418
Total Error	0.0011	1	0.0011		
Total (corr.)	6.1414	7			

R^2 (%): 99.98
Adj R^2 (%): 99.87

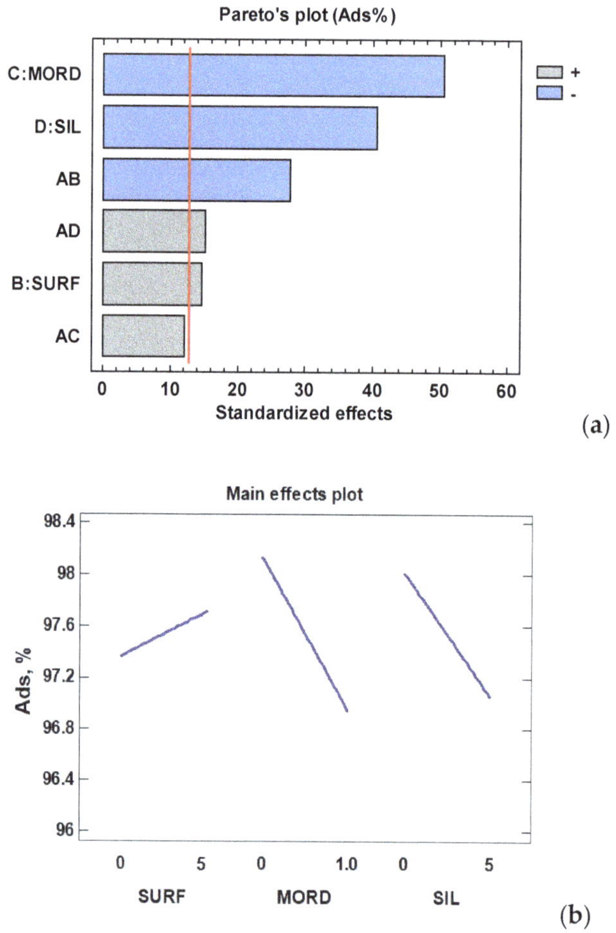

Figure 1. Pareto (**a**) and main effect (**b**) plots for MMT and HT-based chlorophyll hybrid nanopigments.

However, the nanoclay structure used for CD adsorption (HT or MMT) was not found to be significant ($p > 0.05$), and just slightly higher adsorption values were obtained for HT compared to MMT (Table 1). So, the interactions shown between the used modifiers and the nanoclay (Figure 1) should be considered to explain interlayer and surface modifications of the used nanoclays that could favour CD adsorption [41]. The aggregation and molecular orientation of the natural dye and layer charge density are also determinant parameters to be considered in the synthesis process of dye/clay complexes [42]. Additionally, the molecular weight and structure of the natural dyes have been reported to directly affect the dye adsorption process on the absorbent surface. In this sense, lower molecular weight dyes could present a higher mobility during the adsorption process and consequently lead to a higher adsorption rate [43]. In our case, the molecular weight of CD could be difficult to determine, to some extent, due its incorporation into both HT and MMT nanoclays.

Strong significant ($p < 0.05$) interactions between nanoclay-surfactant (negative) and nanoclay-silane (positive) were observed (Figure 2). A positive ($p > 0.05$) interaction between nanoclay-biomordant was also found. The influence of the nanoclay structure and the combined addition of silane, mordant and surfactant modifiers in the nanopigment synthesis of different natural dyes has been previously studied [22]. In this work, the use of the surfactant (SDS) increased the adsorption of the dye into HT, while, in contrast, the presence of CPB retained, to a higher extent, the amount of CD loaded into the clay structure for MMT. Regarding the use of the biomordant and silane, a similar trend was found as the adsorption of CD favoured in the absence of both modifiers. So, it was concluded that for MMT-based CDNPs, it was better not to use any of the selected modifiers, while for HT-based nanopigments, just the addition of the surfactant (CPB) will be needed to increase the CD content adsorbed.

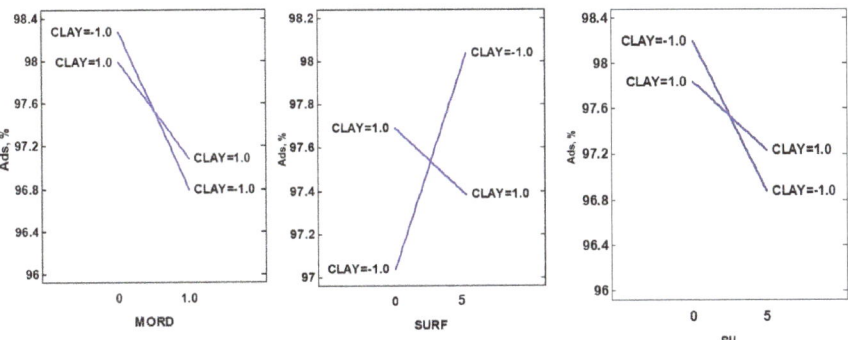

Figure 2. Interaction plots for the synthesis optimization of chlorophyll hybrid nanopigments. For nanoclay codification, (−1) corresponds to HT and (+1) to MMT.

Figure 3a,b show the final appearance and colour of the optimized CDNPs (MMT-C and HT-C, containing, respectively, 0.2 g of chlorophyll dye in 3 g of MMT and HT). In order to obtain a more greenish intense colour, a new sample containing 1 g of chlorophyll dye in 3 g of MMT, without the addition of any modifier, was also obtained (MMT-C_2, Figure 3c).

Figure 3. Visual appearance of chlorophyll hybrid nanopigments obtained with MMT (MMT-C) (a), HT (HT-C) (b) and MMT with a higher chlorophyll dye (CD) concentration (MMT-C_2) (c).

Thermal Stability of CDNPs

The obtained nanopigments (MMT-C, HT-C and MMT-C_2) were also thermally characterized by TGA analysis, and the obtained results are shown in Figure 4. The weight loss profile and the derivative curve of the extracted green chlorophyll dye (CD), without being incorporated in the nanoclays, was also studied and included in the TG/DTG curves. The stability of chlorophyll in plants is known to be affected by a number of factors such as moisture (water), oxygen, temperature, light, pH, long time storage and the presence of enzymes or metals; chlorophyll is easily degraded in the presence of any of these factors resulting in pheophytin as a degradation product in a primary stage. According to Samide and Tutunaru [44], the first step of mass loss in CD observed up to 100 °C was attributed to the removal of physically adsorbed substances on the dye surface during extraction. The chlorophyll transformation reaction continues in the temperature range 100–150 °C, where the decomposition of pheophytin to pyropheophytin takes place. In a next step, the pyropheophytin decomposition to pyropheophorbide began and this step continued up to 500 °C, where the loss was stabilized obtaining a final residue of nearly 20 wt% attributed to carbonaceous magnesium oxide (Figure 4).

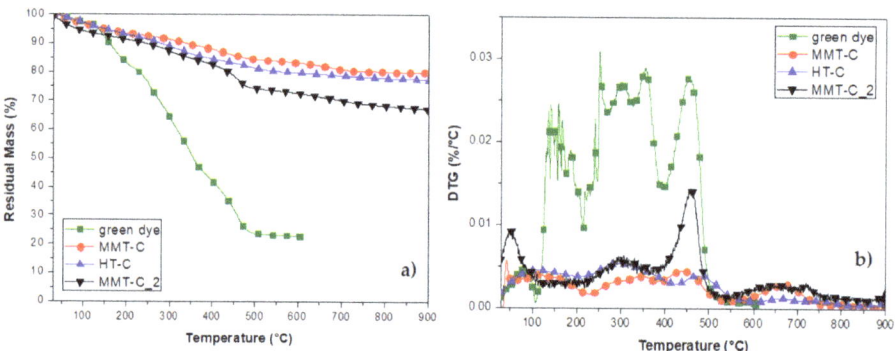

Figure 4. TG (**a**) and DTG (**b**) results obtained for CD (green dye) and chlorophyll hybrid nanopigments (MMT-C, HT-C and MMT-C_2).

The low thermal stability observed for CD was improved by its incorporation in the nanoclays, protecting chlorophyll degradation to some extent under thermal processing conditions (Figure 4). According to reported results and previous observation of unmodified nanoclays, the first step observed in the TG curves of the hybrid nanopigments up to 200 °C was attributed to the desorption of water molecules between the layers, for both MMT and HT; while the further weight loss above 400 °C was ascribed to the dihydroxylation of the remaining OH groups of the nanoclays [8]. The thermal profiles observed for MMT-C and HT-C nanopigments were quite similar, showing a final residual mass of nearly 80 wt% in both dye/clay systems. In addition, the main peaks observed for mass loss in their DTG curves were located at the same temperature range as that observed for CD degradation. On the other hand, the MMT-C_2 nanopigment, with a more brilliant and intense green colour, showed a more evident weight loss and lower thermal stability, essentially due to a higher CD loading on MMT clay, showing a superposition of the thermodegradative behaviour of both MMT and CD components.

3.2. Bionanocomposites Characterization

The produced CDNPs were incorporated at 7 wt% loading for the development and characterization of INZEAF2-based coloured bionanocomposites. Blank samples corresponding to neat INZEA and INZEA containing unmodified MMT and HT at 7 wt% were also obtained. Figure 5 shows the visual images of all obtained biobionanocomposites. A darker green colour was observed for the INZEA_MMT-C_2 sample due to the intrinsic colour of MMT and MMT-C_2 nanopigment which contained a higher CD amount.

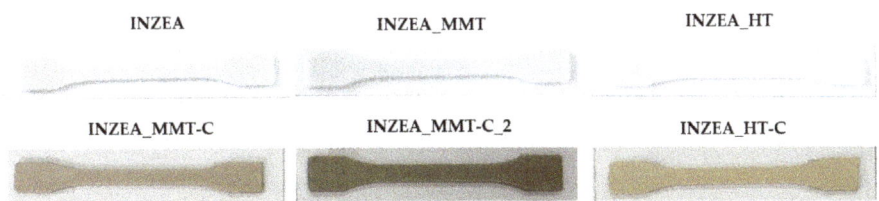

Figure 5. INZEA-based bionanocomposites containing unmodified nanoclays and chlorophyll hybrid nanopigments at 7 wt%.

3.2.1. Evaluation of Colour Parameters

The results obtained for the colour parameters of the developed bionanocomposites are included in Table 3. Neat INZEA was characterized by a high lightness value due to the colour and clear appearance of the neat polymer. The addition of MMT and HT to INZEA-based systems produced, respectively, a reduction and increase in L* values (Table 1). The addition of MMT to INZEA resulted in large ΔE* values compared to HT incorporation, with values of 21.89 and 13.93, respectively. This phenomenon was associated to the MMT and HT intrinsic colours, which were able to modify the final aesthetic quality and appearance of the different samples [8]. The presence of CDNPs in INZEA-based systems determined a remarkable variation in CIELAB values. A visible reduction in L* and a* values and an increase in b* values were observed, compared to the unmodified clays, indicating a deviation of the polymer samples towards green/yellow tones and an overall decrease in the sample lightness. The lower lightness in those samples containing chlorophyll hybrid nanopigments could be attributed to their selective light absorption and green colour which could hinder light transmittance [39]. Regarding MMT-based chlorophyll hybrid nanopigments, the MMT-C sample was characterized by a bright green colour compared to INZEA_MMT-C_2, showing a higher ΔE* value in the latter. Finally, gloss values were reduced by using unmodified nanoclays and partially recovered with the incorporation of CDNPs, which contributed to a great extent to modifying the ΔE* values.

Table 3. CIELAB parameters for INZEA-based bionanocomposites containing 7 wt% of unmodified MMT and HT and chlorophyll hybrid nanopigments (m ± SD; n = 3).

Formulations	L*	a*	b*	Gloss	ΔE*
White Control	99.47 ± 0.00	−0.08 ± 0.01	−0.08 ± 0.01	121	-
INZEA	81.66 ± 0.32	0.65 ± 0.05	5.11 ± 0.04	68 ± 3	18.56 ± 0.30
INZEA_MMT	79.50 ± 0.74	0.90 ± 0.07	8.83 ± 0.16	60 ± 5	21.89 ± 0.66
INZEA_MMT-C	74.22 ± 0.23	0.33 ± 0.04	12.00 ± 0.14	64 ± 1	28.00 ± 0.16
INZEA_MMT-C_2	68.29 ± 0.05	0.13 ± 0.02	12.30 ± 0.03	70 ± 1	33.55 ± 0.04
INZEA_HT	86.22 ± 0.19	0.14 ± 0.04	4.22 ± 0.08	57 ± 1	13.93 ± 0.19
INZEA_HT-C	74.67 ± 0.24	−0.07 ± 0.01	18.71 ± 0.15	65 ± 1	31.12 ± 0.10

3.2.2. Tensile Properties

Tensile characterization results of INZEA-based bionanocomposites are reported in Figure 6. The incorporation of 7 wt% of CDNPs resulted in an increase in Young's modulus of 24.4% and 26.5% for INZEA_MMT-C and INZEA_HT-C, respectively, demonstrating a reinforcement effect of the developed nanopigments in the INZEA matrix. The mechanical properties of the nanocomposites greatly depend on the interaction between the matrix components (polymer and nanoclay). The formation of intercalated/exfoliated structures could enhance the interfacial interactions between the polymer molecular chains and nanopigment layers, resulting in higher stiffness [39]. However, the increase in CD loading in INZEA_MMT-C_2 produced some deterioration in E values compared to INZEA_MMT-C. Marchante et al. [33] also described a similar trend in the Young's modulus when nanoclay-based pigments were incorporated into linear low-density polyethylene. On the other hand, a decrease in terms of strength and strain at break was observed for all formulations, regardless of the type of the nanopigment introduced in the polymer matrix. Similar findings were reported by Esfahani et al. [45] which were attributed to the existence of weak polymer/nanoclay interactions and a decrease in ductility. The formation of some agglomeration of nanopigment particles and the presence of unexfoliated aggregates and structural voids could result in a limited amount of intercalation/exfoliation of MMT and HT in the polymer matrix, affecting the structural stability and mechanical performance of the final bionanocomposites and explaining the observed mechanical results [46].

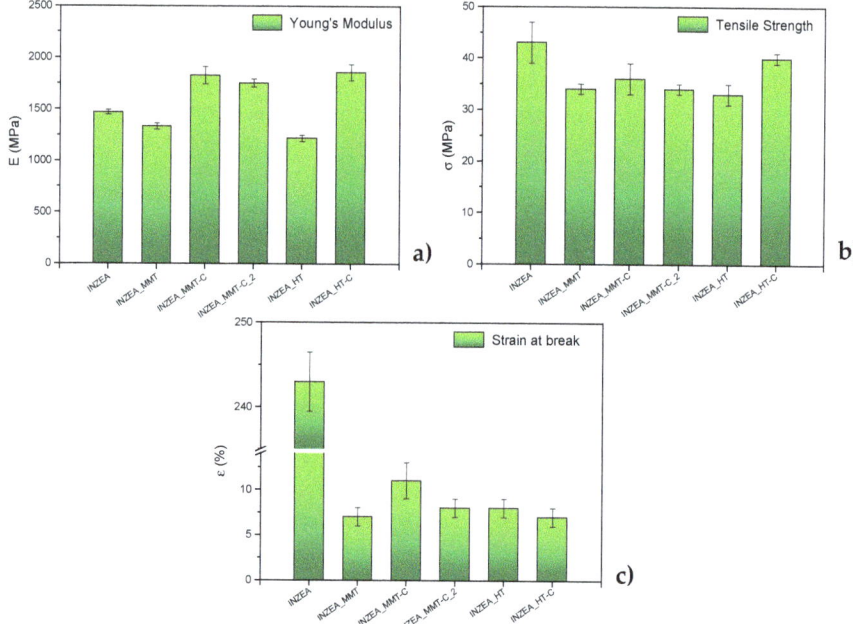

Figure 6. Results of tensile tests for INZEA-based bionanocomposites containing 7 wt% of unmodified MMT and HT and chlorophyll hybrid nanopigments: (a) E, Young's Modulus; (b), σ_b: strength at break and (c) ε_b at σ_b: strain at break. (m ± SD; n = 5).

3.2.3. DMTA Tests

Dynamic mechanical thermal analysis was considered to further evaluate the effect of the addition of the CDNPs on the thermomechanical behaviour of the neat INZEA matrix. The analysis of G' curves (Figure 7a) revealed that while the unmodified MMT and HT incorporation slightly improved the original INZEA values (7.1×10^6 MPa and 7.0×10^6 MPa, respectively, for INZEA_MMT and INZEA_HT compared to 6.1×10^6 MPa for neat INZEA), the storage moduli for the formulations containing hybrid nanofillers were enhanced with respect to the neat matrix. The adsorption of the CDNPs onto the macromolecular chains of the biopolyester led to a constraint in the chains movement, particularly in the case of MMT-based hybrid nanopigments. In particular, after the incorporation of 7 wt% of MMT-C and MMT-C_2, the storage modulus (measured at 40 °C) of neat INZEA increased up to 7.1×10^6 MPa and 7.8×10^6 MPa, respectively for INZEA_MMT-C and INZEA_MMT-C_2, demonstrating the reinforcing effect of high loaded CDNPs in the polyester matrix. The possible formation of an intercalated/exfoliated morphology (as demonstrated in the following section) enhanced the interfacial interactions between the polymer molecular chains and the nanohybrid layers. Consequently, the imposed stress on polymer chains can be effectively suppressed by the rigid and high aspect ratio clay nanosheets of CDNPs [39]. The formation of some aggregates and a decrease in the amount of intercalation/exfoliation of CD in the case of HT-based nanohybrids can explain the observed results in terms of G' values for these hybrid nanopigments, registering storage moduli (measured at 40 °C) of 7.0×10^6 MPa and 7.3×10^6 MPa, respectively, for INZEA_HT and INZEA_HT-C. The upper temperature limit of application of these materials is close to the melting point of the low melting component—i.e., 100 °C—above which the modulus decreases rapidly.

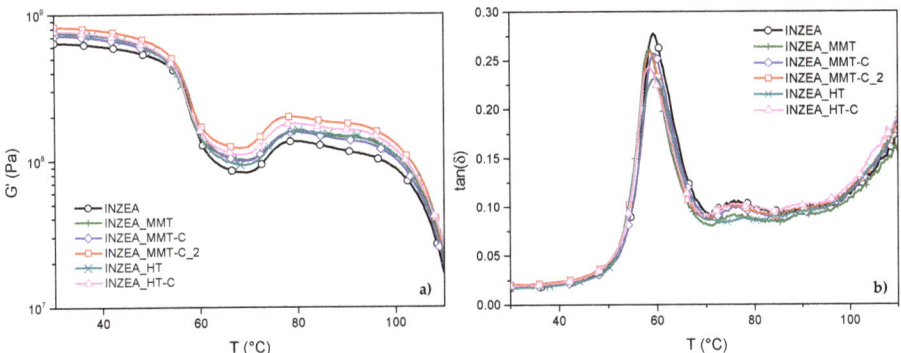

Figure 7. G' (**a**) and tan δ curves (**b**) for INZEA-based bionanocomposites containing 7 wt% of unmodified MMT and HT and chlorophyll hybrid nanopigments.

Regarding the determination of the T_g by DMTA (Figure 7b), a signal related to the presence of a glass transition at around 60 °C was monitored and ascribable to the transition of the high melting fraction; in parallel, a slight variation in glass transition temperatures after the incorporation of the nanofillers for the other samples was noted, meaning that a restriction in the movement of the polyester macromolecular chains occurred, significantly affecting the stiffness of the materials. These results may be explained by the fact that the dispersed nanoclay layers in the polyester matrix can decrease the free volume and hinder the segmental motions of PLA molecular chains at the interface, thus leading to an increase in glass transition temperature [47]. Additionally, a drop in the intensity of the tan δ peak was observed for bionanocomposite samples which was amplified in the presence of CDNPs compared to neat INZEA and filled INZEA with MMT and HT. In this sense, the embedded high aspect ratio CDNP layers in the INZEA matrix could cause a strong interaction between the two components, restricting the polymer chain motions.

3.2.4. Structural and Morphological Properties

Figure 8a shows the XRD patterns of INZEA, INZEA-HT and INZEA-HT-C materials. As can be observed, INZEA showed two sharp and intense peaks at 9.5° and 28.6° that could be attributed to an inorganic filler present in the polymer blend structure. Some authors have attributed these peaks to the crystallographic pattern of talc, in which tetrahedral sheets constituted by a network of silicon and oxygen atoms surrounded by $Mg(OH)_2$ sheets are the main constituents of this mineral reinforcement [48,49]. This result is also in agreement with the TEM micrographs shown in Figure 9 giving a clear indication of the presence of an amount of an inorganic filler in the INZEA structure. The relatively low-intensity peaks around 19–22° match the diffraction of poly(butylene succinate) (PBS), particularly the peaks at $2\theta = 19.4°$ and 22.2°, corresponding to the (020) and (110) crystallographic planes, respectively [50]. Another low-intensity peak was observed at around 16.7°, and it could be attributed to the PLA fraction of the polymer blend, in particular the (200) crystallographic plane, as has already been reported by other authors [51].

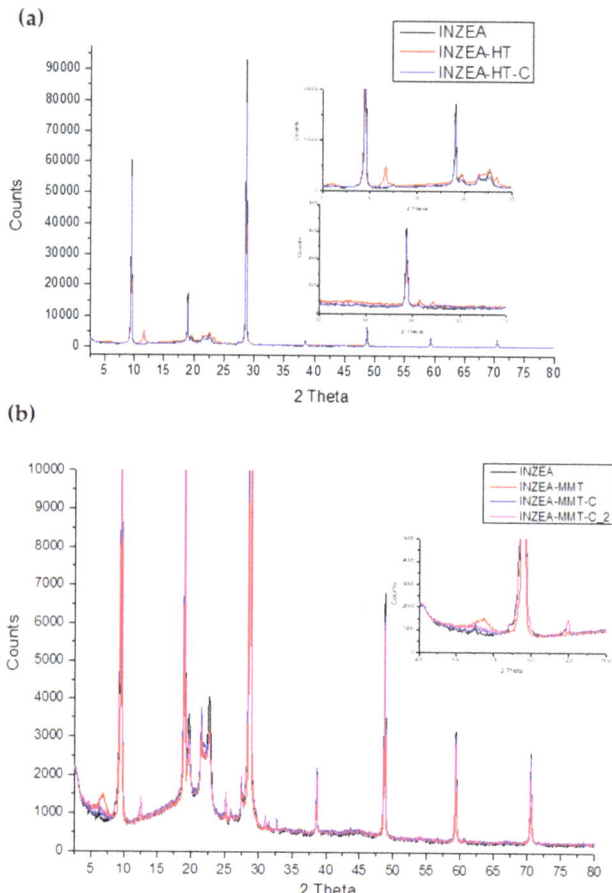

Figure 8. XRD patterns of (**a**) INZEA and INZEA-based bionanocomposites containing 7 wt% of unmodified HT and chlorophyll hybrid nanopigment; (**b**) INZEA and INZEA samples containing 7 wt% of unmodified MMT and chlorophyll hybrid nanopigments.

When HT nanoclay was added to the INZEA matrix, two low-intensity but relevant peaks at around 2θ = 11.6° and 23.4° (see red curve in zoomed area in Figure 8a) were observed. These peaks are indicative of the successful incorporation of this nanoclay into the INZEA structure, and they can be attributed to the (003) and (006) crystallographic planes of calcined HT modified with the addition of a surfactant (SDS), as has already been reported [17,19]. Two other peaks were observed at high angles between 62° and 63° which were assigned to the (110) and (113) crystallographic planes of the surfactant-modified HT, according to previous reported works [19,52,53]. All these peaks assigned to HT were not observed with the incorporation of the chlorophyll hybrid nanopigment (INZEA-HT-C), giving an indication of a modification in the HT crystalline structure and a partial intercalation/exfoliation of CD into the HT laminar structure, resulting in the modification of the HT planes as reported in other formulations [52,53] and in agreement with TEM results (Figure 9).

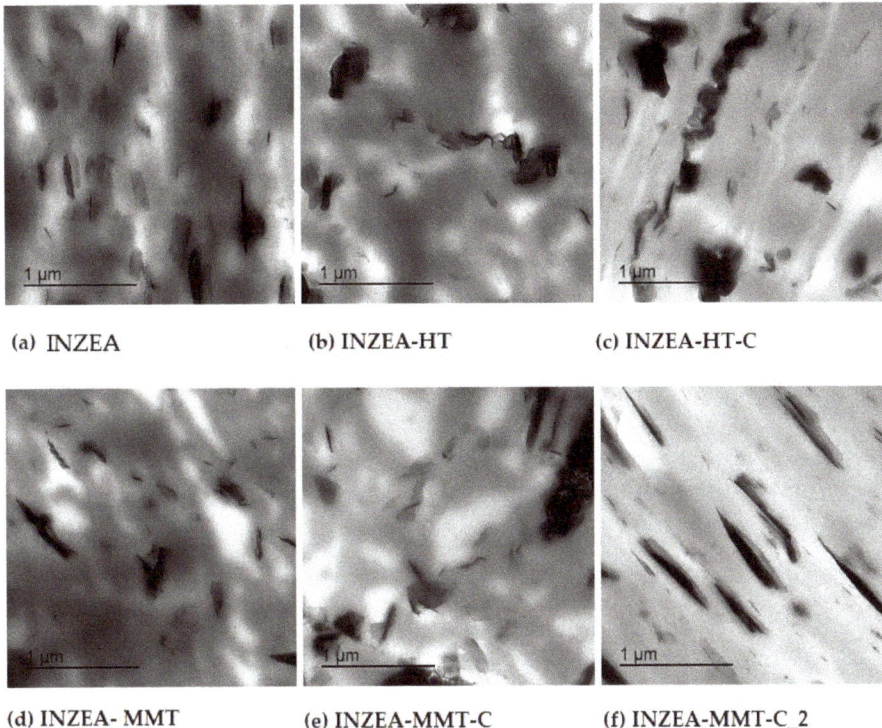

Figure 9. TEM micrographs of INZEA-based bionanocomposites containing 7 wt% of unmodified MMT and HT and chlorophyll hybrid nanopigments.

Figure 8b shows the XRD patterns of INZEA-based formulations modified with MMT and CDNPs at different CD concentrations. The presence of the characteristic crystallographic peak of MMT (around 7°) was confirmed (see red curve in the zoomed area in Figure 8b), corresponding to the (001) plane [19]. The observation of this peak has been extensively reported as indicative of the successful incorporation of MMT into different biopolymer blends, such as those based on PLA and poly(butylene succinate)-co-adipate (PBSA) [54]. However, this peak was not observed when CD was incorporated into the MMT structure giving rise to similar conclusions to the case of HT-based nanocomposites, suggesting a partial intercalation/exfoliation of CD into the MMT laminar structure and the modification of the nanoclay structure with the loss of its crystalline character.

As already stated, the obtained XRD results were in agreement with the TEM micrographs shown in Figure 9. A surface study was performed to precisely evaluate the incorporation and dispersion of CDNPs into the polymer blend structure. The micrograph obtained for INZEA (Figure 9a) showed a random dispersion of tactoids with different sizes all distributed through the polymer matrix. This is another clear indication of the presence of an inorganic filler in the polymer blend structure. The combined XRD and TEM results found for neat INZEA were in agreement with those recently reported by Dias et al. [55], who did not observe a good dispersion and just some homogeneous distribution of the inorganic filler in polyurethane-based nanocomposites. These authors also reported that talc tactoids in these formulations were around 500 nm wide, which are similar to the results obtained in this study.

The addition of hydrotalcite, with or without CD, to the INZEA matrix resulted in the modification of the polymer microstructure, since some agglomeration was observed probably caused by the interaction between both nanofillers (the intrinsic inorganic filler present in the INZEA structure and HT nanoclay), with some indications of a partial intercalation/exfoliation, particularly when CD was present (Figure 9b,c). These results could be explained when considering the anionic character of the surfactant used, since its addition into the HT together with CD would result in the delamination of the nanoclay platelets and the consequent modification of the nanocomposite structure, as observed for other anionic dyes [56]. A similar behaviour was observed for MMT-based bionanocomposites (Figure 9d–f), where the presence of partially intercalated/exfoliated structures after the addition of MMT-based CDNPs was obtained.

However, it should be highlighted that these results were slightly different from those expected, since the intercalation of the used nanoclays, with or without CD addition, should be higher, according to previous results reported for PLA/PBS blends added with organo-modified MMT where a high intercalation/exfoliation degree was reported [57]. The mechanism of incorporation of the nanopigments and the nanoclay itself into the polymer matrix could be clearly interfered by the presence of another filler, as already discussed, making the possibilities for intercalation difficult. These observations suggest that the intercalation of the unmodified MMT and HT and CDNPs into the INZEA structure was more complicated and complex than expected due to the presence of grafted tactoid-like structures in the polymer matrix, as observed with other polymer matrices reinforced with MMT [58]. This effect could also explain the results obtained in the tensile tests which were described in Section 3.2.2.

3.2.5. Thermal stability

A deep analysis of the DTG curves (Figure 10) obtained for bionanocomposites containing HT-based CDNPs clearly indicated that the thermal stability of the overall blend was strongly affected. Specifically, while the temperature for the second main peak remained substantially unaffected, the T_{peak1} for the less stable component in the INZEA matrix was shifted from 350 to 291 °C for INZEA_HT-C. An analogous behaviour, even if limited in the shift values, was noted for MMT-based materials, where T_{peak1} for the less stable component (registered at 352 °C) shifted to 346 and 337 °C for INZEA_MMT-C and INZEA_MMT-C_2, respectively. According to the literature, this degradation path can be rationalized by considering the organic nature of the nanopigments, which were indeed responsible for the decrease in thermal stability of the polyester-based matrix, due to a possible hydrolysis reaction [59,60]. On the other hand, a residual mass for the unfilled matrix of 4 wt% was observed at 900 °C indicating the presence of an inorganic filler in the formulation of the commercial product, as already reported in a previous study [61] and observed in XRD and TEM results. The incorporation of CDNPs into the polymer matrix limited the movement of the polymer phase, thereby reducing the residual weight loss of the bionanocomposite materials, with a mean increase of 13 wt%. It is well known that only when nanoclay layers are exfoliated, they may have reduced the volatilization of the degradation products, and then, the thermal stability could increase [62]. In our case, the already observed limited mechanical enhancement, due to reduced exfoliation of the nanoclays, can also justify the reduced thermal stability to some extent in the obtained formulations.

Figure 11a,b show the DSC thermograms obtained for bionanocomposites during the cooling and second heating scans. Two different peaks were observed for neat INZEA, indicating the presence of the two main polyesters in the polymer matrix, in agreement with the behaviour previously observed in a previous work [8]. While similar values for glass transition temperatures were registered, a significant effect due to the presence of CDNPs on the melting/crystallization behaviour of the polymer matrix was detected. A different trend was noted in the shift of T_c temperatures in the case of low and high melting fractions, and at the same time MMT- and HT-based nanopigments induced a different crystallization effect (Figure 11a). In the case of MMT-based formulations (INZEA_MMT-C and INZEA_MMT-C_2), a slight shift towards higher melting temperatures (from 115 °C for the neat polymer to 117 °C for bionanocomposites) was observed, while the crystallization was indeed delayed observing a peak shift to lower temperatures (from 86 °C for neat INZEA to 83 and 81 °C, respectively, for INZEA_MMT-C and INZEA_MMT-C_2). By considering the possible composition of the commercial INZEA as a blend of two polyesters, a better dispersion for the MMT-based CDNPs in the lower melting phase was suggested, according to the observed melting curves (second scan, Figure 11b), which registered small melting variations only for this fraction. On the other hand, it was found that the presence of nanopigments intercalated in HT altered the bimodal melting behaviour of the high melting fraction, and it is expected that a concurrent effect occurred, for a limited nucleating action and partial hydrolysis of the polyester agent due to the presence of hydroxyl surface functionalities for HT [63].

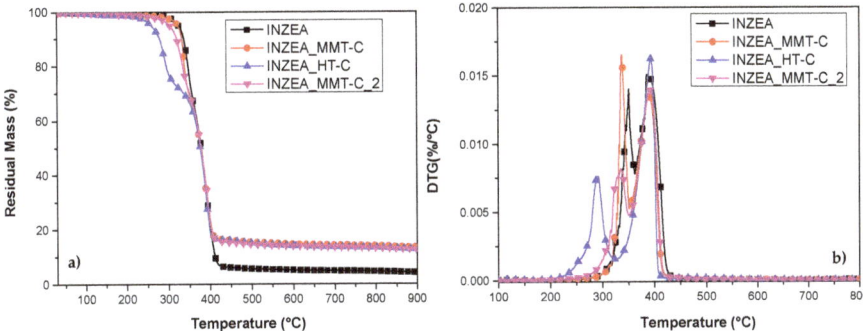

Figure 10. TG (**a**) and DTG (**b**) profiles obtained for INZEA-based bionanocomposites containing 7 wt% of functionalized MMT and HT with CD.

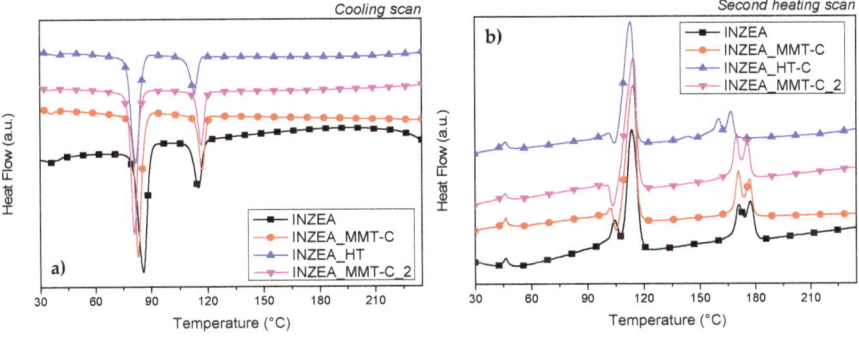

Figure 11. Curves of DSC cooling (**a**) and second heating (**b**) scans performed at 10 °C min^{-1} for INZEA-based bionanocomposites containing 7 wt% of functionalized MMT and HT with CD.

4. Conclusions

Coloured reinforced INZEA-based bionanocomposites containing 7 wt % of chlorophyll hybrid nanopigments (CDNPs) were developed by melt blending. CDNPs were successfully synthetized by using a chlorophyll dye (CD) obtained from broccoli wastes and two nanoclays, MMT and calcined HT. The best synthesis conditions for CD adsorption were studied by using a 2^{4-1} fractional factorial design of experiments. The solvent used was 50% (v/v) ethanol, and the total amount of dye loaded was 0.2 g in 3 g of nanoclay. The incorporation of CD into the nanoclays improved its thermal stability. The maximum CD adsorption into MMT was obtained without the addition of any modifier, while 5 wt% of surfactant (sodium dodecyl sulphate) was needed in the case of HT. Interesting green/yellow tones were obtained with the addition of CDNPs to the INZEA polyester matrix. The morphological and structural XRD characterization of INZEA-based bionanocomposites showed the coexistence of an inorganic filler in the commercial formulation together with the hybrid nanofillers, affecting the agglomeration and intercalation/exfoliation state of CDNPs in the polymer matrix. For the same reason, the incorporation of CDNPs into the polymer matrix was found to affect the thermal stability of the bionanocomposites, by reducing, to some extent, the temperature of maximum degradation rate for the less thermally stable component in the INZEA matrix. The melting/crystallization behaviour of the polyester-based matrix was also modified by the presence of the CDNPs, since the presence of the nanopigments altered the bimodal melting behaviour of the high melting fraction and delayed the crystallization towards lower temperatures.

The developed coloured polyester-based bionanocomposites could be considered as promising functional materials to be potentially used in different applications where desired colour and reinforcement properties could be an issue, such as in the automotive sector. In addition, the use of chlorophyll dyes obtained from broccoli wastes could contribute to the valorisation of these residues and the general objective of the circular economy approach. Further work will be needed to improve the interaction between the polymer matrix and the nanopigments to enhance the reinforcement properties of the obtained biomaterials for other potential applications.

Author Contributions: Conceptualization, A.J., M.C.G., V.V., L.T. and D.P.; methodology B.M.-V., M.R., F.D., and F.L.; validation, A.J., M.C.G., L.T. and D.P.; formal analysis, B.M.-V., M.R., F.D., and F.L.; investigation, B.M., M.R., F.D., and F.L.; resources, A.J., M.C.G., L.T. and D.P.; data curation, B.M.-V., M.R., F.D., F.L., V.V., A.J., M.C.G. and D.P.; writing—original draft preparation, B.M.-V., M.R., F.D., and F.L.; writing—review and editing, A.J., M.C.G. and D.P.; supervision, A.J., M.C.G., V.V., L.T. and D.P.; funding acquisition, A.J., M.C.G., L.T. and D.P. All authors have read and agreed to the published version of the manuscript.

Funding: The authors express their gratitude to the Bio Based Industries Consortium and European Commission for the financial support to the project BARBARA: Biopolymers with advanced functionalities for building and automotive parts processed through additive manufacturing. This project received funding from the Bio Based Industries Joint Undertaking under the European Union's Horizon 2020 research and innovation programme under grant agreement No 745578.

Conflicts of Interest: The authors declare no conflict of interest.

References

1. Ebrahimi, I.; Parvinzadeh Gashti, M. Extraction of polyphenolic dyes from henna, pomegranate rind, and *Pterocarya fraxinifolia* for nylon 6 dyeing. *Color. Technol.* **2016**, *132*, 162–176. [CrossRef]
2. Shalini, S.; Prasanna, S.; Mallick, T.K.; Senthilarasu, S. Review on natural dye sensitized solar cells: Operation, materials and methods. *Renew. Sustain. Energy Rev.* **2015**, *51*, 1306–1325. [CrossRef]
3. Heinonen, J.; Farahmandazad, H.; Vuorinen, A.; Kallio, H.; Yang, B.; Sainio, T. Extraction and purification of anthocyanins from purple-fleshed potato. *Food Bioprod. Process.* **2016**, *99*, 136–146. [CrossRef]
4. Keppler, K.; Humpf, H.-U. Metabolism of anthocyanins and their phenolic degradation products by the intestinal microflora. *Bioorg. Med. Chem.* **2005**, *13*, 5195–5205. [CrossRef] [PubMed]
5. Ferreira, S.S.; Monteiro, F.; Passos, C.P.; Silva, A.M.S.; Wessel, D.F.; Coimbra, M.A.; Cardoso, S.M. Blanching impact on pigments, glucosinolates, and phenolics of dehydrated broccoli by-products. *Food Res. Int.* **2020**, *132*, 109055. [CrossRef] [PubMed]
6. Thomas, M.; Badr, A.; Desjardins, Y.; Gosselin, A.; Angers, P. Characterization of industrial broccoli discards (*Brassica oleracea var. italica*) for their glucosinolate, polyphenol and flavonoid contents using UPLC MS/MS and spectrophotometric methods. *Food Chem.* **2018**, *245*, 1204–1211. [CrossRef] [PubMed]
7. Ferreira, S.S.; Passos, C.P.; Cardoso, S.M.; Wessel, D.F.; Coimbra, M.A. Microwave assisted dehydration of broccoli by-products and simultaneous extraction of bioactive compounds. *Food Chem.* **2018**, *246*, 386–393. [CrossRef]
8. Micó-Vicent, B.; Viqueira, V.; Ramos, M.; Luzi, F.; Dominici, F.; Torre, L.; Jiménez, A.; Puglia, D.; Garrigós, M.C. Effect of Lemon Waste Natural Dye and Essential Oil Loaded into Laminar Nanoclays on Thermomechanical and Color Properties of Polyester Based Bionanocomposites. *Polymers* **2020**, *12*, 1451. [CrossRef]
9. Dhafina, W.A.; Daud, M.Z.; Salleh, H. The sensitization effect of anthocyanin and chlorophyll dyes on optical and photovoltaic properties of zinc oxide based dye-sensitized solar cells. *Optik* **2019**, *207*, 163808. [CrossRef]
10. Nan, H.; Shen, H.-P.; Wang, G.; Xie, S.-D.; Yang, G.-J.; Lin, H. Studies on the optical and photoelectric properties of anthocyanin and chlorophyll as natural co-sensitizers in dye sensitized solar cell. *Opt. Mater.* **2017**, *73*, 172–178. [CrossRef]

11. Steet, J.A.; Tong, C.H. Degradation kinetics of green color and chlorophylls in peas by colorimetry and HPLC. *J. Food Sci.* **1996**, *61*, 924–928. [CrossRef]
12. Ngamwonglumlert, L.; Devahastin, S.; Chiewchan, N. Natural colorants: Pigment stability and extraction yield enhancement via utilization of appropriate pretreatment and extraction methods. *Crit. Rev. Food Sci. Nutr.* **2017**, *57*, 3243–3259. [CrossRef]
13. Zhang, R.; Zhou, L.; Li, J.; Oliveira, H.; Yang, N.; Jin, W.; Zhu, Z.; Li, S.; He, J. Microencapsulation of anthocyanins extracted from grape skin by emulsification/internal gelation followed by spray/freeze-drying techniques: Characterization, stability and bioaccessibility. *LWT* **2020**, *123*, 109097. [CrossRef]
14. Hsiao, C.J.; Lin, J.F.; Wen, H.Y.; Lin, Y.M.; Yang, C.H.; Huang, K.S.; Shaw, J.F. Enhancement of the stability of chlorophyll using chlorophyll-encapsulated polycaprolactone microparticles based on droplet microfluidics. *Food Chem.* **2020**, *306*, 125300. [CrossRef] [PubMed]
15. Bertuoli, P.T.; Piazza, D.; Scienza, L.C.; Zattera, A.J. Preparation and characterization of montmorillonite modified with 3-aminopropyltriethoxysilane. *Appl. Clay Sci.* **2014**, *87*, 46–51. [CrossRef]
16. Beltrán, M.I.; Benavente, V.; Marchante, V.; Dema, H.; Marcilla, A. Characterisation of montmorillonites simultaneously modified with an organic dye and an ammonium salt at different dye/salt ratios. Properties of these modified montmorillonites EVA nanocomposites. *Appl. Clay Sci.* **2014**, *97*, 43–52. [CrossRef]
17. Kohno, Y.; Asai, S.; Shibata, M.; Fukuhara, C.; Maeda, Y.; Tomita, Y.; Kobayashi, K. Improved photostability of hydrophobic natural dye incorporated in organo-modified hydrotalcite. *J. Phys. Chem. Solids* **2014**, *75*, 945–950. [CrossRef]
18. Su, L.; Tao, Q.; He, H.; Zhu, J.; Yuan, P.; Zhu, R. Silylation of montmorillonite surfaces: Dependence on solvent nature. *J. Colloid Interface Sci.* **2013**, *391*, 16–20. [CrossRef] [PubMed]
19. Micó-Vicent, B.; Jordán, J.; Perales, E.; Martínez-Verdú, F.M.; Cases, F. Finding the Additives Incorporation Moment in Hybrid Natural Pigments Synthesis to Improve Bioresin Properties. *Coatings* **2019**, *9*, 34. [CrossRef]
20. Rather, L.J.; Shabbir, M.; Bukhari, M.N.; Shahid, M.; Khan, M.A.; Mohammad, F. Ecological dyeing of Woolen yarn with Adhatoda vasica natural dye in the presence of biomordants as an alternative copartner to metal mordants. *J. Environ. Chem. Eng.* **2016**, *4*, 3041–3049. [CrossRef]
21. Wu, S.; Cui, H.; Wang, C.; Hao, F.; Liu, P.; Xiong, W. In situ self-assembled preparation of the hybrid nanopigment from raw sepiolite with excellent stability and optical performance. *Appl. Clay Sci.* **2018**, *163*, 1–9. [CrossRef]
22. Micó-Vicent, B.; Jordán, J.; Martínez-Verdú, F.; Balart, R. A combination of three surface modifiers for the optimal generation and application of natural hybrid nanopigments in a biodegradable resin. *J. Mater. Sci.* **2017**, *52*, 889–898. [CrossRef]
23. Fournier, F.; de Viguerie, L.; Balme, S.; Janot, J.-M.; Walter, P.; Jaber, M. Physico-chemical characterization of lake pigments based on montmorillonite and carminic acid. *Appl. Clay Sci.* **2016**, *130*, 12–17. [CrossRef]
24. Armağan, B.; Özdemir, O.; Turan, M.; Celik, M.S. The removal of reactive azo dyes by natural and modified zeolites. *J. Chem. Technol. Biotechnol. Int. Res. Process. Environ. Clean Technol.* **2003**, *78*, 725–732. [CrossRef]
25. Raha, S.; Quazi, N.; Ivanov, I.; Bhattacharya, S. Dye/Clay intercalated nanopigments using commercially available non-ionic dye. *Dye. Pigment.* **2012**, *93*, 1512–1518. [CrossRef]
26. Sommer Márquez, A.E.; Lerner, D.A.; Fetter, G.; Bosch, P.; Tichit, D.; Palomares, E. Preparation of layered double hydroxide/chlorophyll a hybrid nano-antennae: A key step. *Dalt. Trans.* **2014**, *43*, 10521. [CrossRef] [PubMed]
27. Campanholi, K.D.S.S.; Braga, G.; Da Silva, J.B.; Da Rocha, N.L.; De Francisco, L.M.B.; De Oliveira, É.L.; Bruschi, M.L.; De Castro-Hoshino, L.V.; Sato, F.; Hioka, N.; et al. Biomedical Platform Development of a Chlorophyll-Based Extract for Topic Photodynamic Therapy: Mechanical and Spectroscopic Properties. *Langmuir* **2018**, *34*, 8230–8244. [CrossRef]
28. Arof, A.K.; Ping, T.L. Chlorophyll as Photosensitizer in Dye-Sensitized Solar Cells. In *Chlorophyll*; InTech: London, UK, 2017.
29. Ahmed, J.K. Effect of Chlorophyll and Anthocyanin on the Secondary Bonds of Poly Vinyl Chloride (PVC). *Int. J. Mater. Sci. Appl.* **2015**, *4*, 21. [CrossRef]
30. Chandrappa, H.; Bhajantri, R.F.; Prarthana, N. Simple fabrication of PVA-ATE (Amaranthus tricolor leaves extract) polymer biocomposites: An efficient UV-Shielding material for organisms in terrestrial and aquatic ecosystems. *Opt. Mater.* **2020**, *109*, 110204.

31. Chavoshizadeh, S.; Pirsa, S.; Mohtarami, F. Conducting/smart color film based on wheat gluten/chlorophyll/polypyrrole nanocomposite. *Food Packag. Shelf Life* **2020**, *24*, 100501. [CrossRef]
32. Chavoshizadeh, S.; Pirsa, S.; Mohtarami, F. Sesame oil oxidation control by active and smart packaging system using wheat gluten/chlorophyll film to increase shelf life and detecting expiration date. *Eur. J. Lipid Sci. Technol.* **2020**, *122*, 1900385. [CrossRef]
33. Marchante, V.; Marcilla, A.; Benavente, V.; Martínez-Verdú, F.M.; Beltrán, M.I. Linear low-density polyethylene colored with a nanoclay-based pigment: Morphology and mechanical, thermal, and colorimetric properties. *J. Appl. Polym. Sci.* **2013**, *129*, 2716–2726. [CrossRef]
34. Marchante, V.; Benavente, V.; Marcilla, A.; Martínez-Verdú, F.M.; Beltrán, M.I. Ethylene vinyl acetate/nanoclay-based pigment composites: Morphology, rheology, and mechanical, thermal, and colorimetric properties. *J. Appl. Polym. Sci.* **2013**, *130*, 2987–2994. [CrossRef]
35. Mahmoodi, A.; Ebrahimi, M.; Khosravi, A.; Eivaz Mohammadloo, H. A hybrid dye-clay nano-pigment: Synthesis, characterization and application in organic coatings. *Dye. Pigment.* **2017**, *147*, 234–240. [CrossRef]
36. León-Cabezas, M.A.; Martínez-García, A.; Varela-Gandía, F.J. Innovative functionalized monofilaments for 3D printing using fused deposition modeling for the toy industry. *Procedia Manuf.* **2017**, *13*, 738–745. [CrossRef]
37. ColorFabb and Grafe collaborate to introduce colour on demand for 3D printing filaments. *Addit. Polym.* **2018**, *2018*, 4. [CrossRef]
38. Lee, J.-Y.; An, J.; Chua, C.K. Fundamentals and applications of 3D printing for novel materials. *Appl. Mater. Today* **2017**, *7*, 120–133. [CrossRef]
39. Mahmoodi, A.; Ghodrati, S.; Khorasani, M. High-strength, low-permeable, and light-protective nanocomposite films based on a hybrid nanopigment and biodegradable PLA for food packaging applications. *ACS Omega* **2019**, *4*, 14947–14954. [CrossRef]
40. Usop, R.; Abidin, Z.H.Z.; Mazni, N.A.; Hadi, A.N.; Halim, N.A.; Taha, R.M.; Careem, M.A.; Majid, S.R.; Arof, A.K. The colour stability of natural dye coating films consisting of chlorophyll after exposed to UV-A. *Pigment Resin Technol.* **2016**, *45*, 149–157. [CrossRef]
41. Kohno, Y.; Inagawa, M.; Ikoma, S.; Shibata, M.; Matsushima, R.; Fukuhara, C.; Tomita, Y.; Maeda, Y.; Kobayashi, K. Stabilization of a hydrophobic natural dye by intercalation into organo-montmorillonite. *Appl. Clay Sci.* **2011**, *54*, 202–205. [CrossRef]
42. Kaneko, Y.; Iyi, N.; Bujdák, J.; Sasai, R.; Fujita, T. Effect of layer charge density on orientation and aggregation of a cationic laser dye incorporated in the interlayer space of montmorillonites. *J. Colloid Interface Sci.* **2004**, *269*, 22–25. [CrossRef]
43. Bahrami, S.H.; Movassagh, B.; Amirshahi, S.H.; Arami, M. Removal of disperse blue 56 and disperse red 135 dyes from aqueous dispersions by modified montmorillonite nanoclay. *Chem. Ind. Chem. Eng. Q* **2017**, *23*, 21–29.
44. Samide, A.; Tutunaru, B. Thermal behavior of the chlorophyll extract from a mixture of plants and seaweed. *J. Therm. Anal. Calorim.* **2017**, *127*, 597–604. [CrossRef]
45. Esfahani, J.M.; Sabet, A.R.; Esfandeh, M. Assessment of nanocomposites based on unsaturated polyester resin/nanoclay under impact loading. *Polym. Adv. Technol.* **2012**, *23*, 817–824. [CrossRef]
46. Zdiri, K.; Elamri, A.; Hamdaoui, M.; Harzallah, O.; Khenoussi, N.; Brendlé, J. Reinforcement of recycled PP polymers by nanoparticles incorporation. *Green Chem. Lett. Rev.* **2018**, *11*, 296–311. [CrossRef]
47. Darie, R.N.; Pâslaru, E.; Sdrobis, A.; Pricope, G.M.; Hitruc, G.E.; Poiată, A.; Baklavaridis, A.; Vasile, C. Effect of nanoclay hydrophilicity on the poly(lactic acid)/clay nanocomposites properties. *Ind. Eng. Chem. Res.* **2014**, *53*, 7877–7890. [CrossRef]
48. Świetlicki, M.; Chocyk, D.; Klepka, T.; Prószyński, A.; Kwaśniewska, A.; Borc, J.; Gładyszewski, G. The structure and mechanical properties of the surface layer of polypropylene polymers with talc additions. *Materials.* **2020**, *13*, 698. [CrossRef]
49. Pi-Puig, T.; Animas-Torices, D.Y.; Solé, J. Mineralogical and geochemical characterization of talc from two mexican ore deposits (oaxaca and puebla) and nine talcs marketed in mexico: Evaluation of its cosmetic uses. *Minerals* **2020**, *10*, 388. [CrossRef]
50. Wang, K.; Jiao, T.; Wang, Y.; Li, M.; Li, Q.; Shen, C. The microstructures of extrusion cast biodegradable poly(butylene succinate) films investigated by X-ray diffraction. *Mater. Lett.* **2013**, *92*, 334–337. [CrossRef]

51. Zhang, H.; Wang, S.; Zhang, S.; Ma, R.; Wang, Y.; Cao, W.; Liu, C.; Shen, C. Crystallization behavior of poly(lactic acid) with a self-assembly aryl amide nucleating agent probed by real-time infrared spectroscopy and X-ray diffraction. *Polym. Test.* **2017**, *64*, 12–19. [CrossRef]
52. Heraldy, E.; Nugrahaningtyas, K.D. X-RAY diffraction and fourier transform infrared study of Ca-Mg-Al hydrotalcite from artificial brine water with synthesis hydrothermal treatments. *IOP Conf. Ser. Mater. Sci. Eng. Pap.* **2018**, *333*.
53. Bascialla, G.; Regazzoni, A.E. Immobilization of anionic dyes by intercalation into hydrotalcite. *Colloids Surfaces A Physicochem. Eng. Asp.* **2008**, *328*, 34–39. [CrossRef]
54. Abdallah, W.; Mirzadeh, A.; Tan, V.; Kamal, M.R. Influence of nanoparticle pretreatment on the thermal, rheological and mechanical properties of PLA-PBSA nanocomposites incorporating cellulose nanocrystals or montmorillonite. *Nanomaterials* **2019**, *9*, 29. [CrossRef]
55. Dias, G.; Prado, M.; Le Roux, C.; Poirier, M.; Micoud, P.; Ligabue, R.; Martin, F.; Einloft, S. Synthetic talc as catalyst and filler for waterborne polyurethane-based nanocomposite synthesis. *Polym. Bull.* **2020**, *77*, 975–987. [CrossRef]
56. Costa, A.L.; Gomes, A.C.; Pillinger, M.; Gonçalves, I.S.; de Melo, J.S.S. An indigo carmine-based hybrid nanocomposite with supramolecular control of dye aggregation and photobehavior. *Chem. Eur. J.* **2015**, *21*, 12069–12078. [CrossRef] [PubMed]
57. Tan, L.; He, Y.; Qu, J. Structure and properties of polylactide/poly(butylene succinate)/organically modified montmorillonite nanocomposites with high-efficiency intercalation and exfoliation effect manufactured via volume pulsating elongation flow. *Polymer* **2019**, *180*, 121656. [CrossRef]
58. Tanna, V.A.; Enokida, J.S.; Coughlin, E.B.; Winter, H.H. Functionalized polybutadiene for clay-polymer nanocomposite fabrication. *Macromolecules* **2019**, *52*, 6135–6141. [CrossRef]
59. Delpouve, N.; Saiter-Fourcin, A.; Coiai, S.; Cicogna, F.; Spiniello, R.; Oberhauser, W.; Legnaioli, S.; Ishak, R.; Passaglia, E. Effects of organo-LDH dispersion on thermal stability, crystallinity and mechanical features of PLA. *Polymer* **2020**, *208*, 122952. [CrossRef]
60. Coiai, S.; Cicogna, F.; de Santi, A.; Pérez Amaro, L.; Spiniello, R.; Signori, F.; Fiori, S.; Oberhauser, W.; Passaglia, E. MMT and LDH organo-modification with surfactants tailored for PLA nanocomposites. *Express Polym. Lett.* **2017**, *11*, 163–175. [CrossRef]
61. Ramos, M.; Dominici, F.; Luzi, F.; Jimenez, A.; Garrigós, M.; Torre, L.; Puglia, D. Effect of almond shell waste on physicochemical properties of polyester-based biocomposites. *Polymers* **2020**, *12*, 835. [CrossRef]
62. Bikiaris, D. Can nanoparticles really enhance thermal stability of polymers? Part II: An overview on thermal decomposition of polycondensation polymers. *Thermochim. Acta* **2011**, *523*, 25–45. [CrossRef]
63. Mohapatra, A.K.; Mohanty, S.; Nayak, S.K. Study of thermo-mechanical and morphological behaviour of biodegradable PLA/PBAT/layered silicate blend nanocomposites. *J. Polym. Environ.* **2014**, *22*, 398–408. [CrossRef]

Publisher's Note: MDPI stays neutral with regard to jurisdictional claims in published maps and institutional affiliations.

© 2020 by the authors. Licensee MDPI, Basel, Switzerland. This article is an open access article distributed under the terms and conditions of the Creative Commons Attribution (CC BY) license (http://creativecommons.org/licenses/by/4.0/).

Article

On the Combined Effect of Both the Reinforcement and a Waste Based Interfacial Modifier on the Matrix Glass Transition in iPP/a-PP-*p*PBMA/Mica Composites

Jesús María García Martínez * and Emilia P. Collar *

Polymer Engineering Group (GIP), Polymer Science and Technology Institute (ICTP), Spanish Council for Scientific Research (CSIC), C/Juan de la Cierva, 3, 28006 Madrid, Spain
* Correspondence: jesus.maria@ictp.csic.es (J.M.G.M.); ecollar@ictp.csic.es (E.P.C.)

Received: 30 September 2020; Accepted: 3 November 2020; Published: 6 November 2020

Abstract: This work deals with the changes of the glass transition temperature (T_g) of the polymer in polypropylene/mica composites due to the combined and synergistic effect of the reinforcement and the interfacial modifier. In our case, we studied the effect on T_g of platy mica and an interfacial modifier with *p*-phenylen-bis-maleamic acid (*p*PBMA) grafted groups onto atactic polypropylene (aPP-*p*PBMA). This one contains 5.0×10^{-4} g·mol^{-1} (15% *w/w*) grafted *p*PBMA and was previously obtained by the author's labs by using industrial polymerization wastes (aPP). The objective of the article must be perceived as two-fold. On one hand, the determination of the changes in the glass transition temperature of the isotactic polypropylene phase (iPP) due to both the reinforcement and the agent as determined form the damp factor in DMA analysis. On the other hand, forecasting the variation of this parameter (T_g) as a function of both the interfacial agent and reinforcement content. For such purposes, and by assuming the complex character of the iPP/aPP-*p*PBMA/Mica system, wherein interaction between the components will define the final behaviour, a Box–Wilson experimental design considering the amount of mica particles and of interface agent as the independent variables, and the T_g as the dependent one, has been used. By taking in mind that the glass transition is a design threshold for the ultimate properties of parts based in this type of organic–inorganic hybrid materials, the final purpose of the work is the prediction and interpretation of the effect of both variables on this key parameter.

Keywords: organic–inorganic hybrid materials; compatibilizers; composites; modeling; interfaces; wastes; residues; iPP; aPP

1. Introduction

The International Union of Pure and Applied Chemistry (IUPAC) defines nano-composite as that composite in which at least one of the phase domains has at least one dimension of the order of nanometers [1]. Furthermore, it defines a hybrid material as the one composed of an intimate mixture of inorganic components, organic components, or both types of components that usually interpenetrate on scales of less than 1 μm [1]. A detailed and comprehensive description and recommendations for studying this kind of system can found elsewhere [2]. The above-mentioned definitions match the iPP/mica composites, wherein the mica reinforcement has one dimension (thickness) in the nano-scale order close to 30 nm even if non-exfoliated [3,4]. In the same way, organic–inorganic hybrid material. In the same way, organic–inorganic materials can be revealed as multi-component compounds having at least one of their organic (the polymer) or inorganic component in the sub-micrometric and more usually in the nano-metric size domain [2,5]. It is worth it to mention that mica is a well-recognized mineral by its virtually perfect cleavage capacity at the atomic range level. It explains the excellent dimensional stability of mica/polymer composites in addition to excellent surface quality. Decidedly, the high-energy dissipation capabilities of mica confer the material a significant vibration and sound suppression. This fact makes it useful in many industries such as welding electrodes manufacture, gypsum plasterboards, paints, and rubber and plastics compounds [3,6]. Under these premises, the use of mica as the inorganic phase represents a wonderfully attractive way for obtaining iPP based organic–inorganic hybrid materials due to the fact that this mineral can be easily cleaved and delaminated in ultra-thin flakes with very high aspect ratios, making easy the alignment of the flakes in the matrix during the processing operations, providing high reinforcement level [6].

It is also completely accepted that organic–inorganic hybrid systems can be classified into two classes named Class I and Class II [5]. Class I organic–inorganic hybrid materials implies that the interaction between the phases is weak and principally due to Van der Waals, hydrogen, electrostatic, and so on, interactions [2,5]. On the contrary, Class II hybrids imply that the interaction between phases is intense due to real chemical bonds between them. The coexistence of both types of interactions (I and II) in the same system is also possible [2,5].

The above-mentioned atomic cleavage capacity represents a feature that confers almost total hydrophobicity to the mica mineral surfaces. The latter prevents the adverse effect at the mineral/matrix interface by the atmosphere water typically anchored to other silicate-based minerals. It explains the high dissipation capabilities below the glass transition temperature of the matrix in polypropylene/mica composites once the percolation threshold between the mica particles is reached [3,6]. Regardless, the affinity lack between the non-polar matrix and the polar mica particles still remains. Therefore, the employment of interfacial agents (or compatibilizers) has demonstrated an ability to enhance the interactions between the components in heterogeneous hybrid materials wherein one of the phases is a polymer [7–13]. Thus, it becomes desirable that the interfacial agent must resemble chemically the polymer phase besides demonstrating an affinity with the reinforcement. That means that in absence of an interfacial agent (due to the fact that mica and polypropylene exhibit very different polarity), the interface interaction level is relatively poor, and so, the inter-phase between components is weak. Consequently, the composite becomes a Class I organic–inorganic hybrid material. At this point, in spite of the research efforts performed, there is still much work to do for enhancing and precisely interpreting the complex phenomena taking place [2–4,11–16].

It is well worth it to establish that the remoter properties of the hybrid material strongly depend on the type and amount of interfacial agent used [7,11–15]. Besides, it is worth it to mention that usually just a scant amount of the interfacial agent is enough and mandatory to optimize the whole behavior of the hybrid material. Therefore, a critical amount of it jointly with those of the other components must emerge [3,4,7,8,15]. In addition, the authors reported the high dissipation capability of the matrix (iPP) below its glass transition when the percolation threshold between the mica particles is reached. The latter was performed for both injected and compressed composites by using DMA spectroscopy [3,8,13]. The iPP/aPP-pPBMA/Mica has been already characterized in previous works by tensile, flexural, and impact properties. Additionally, the crystalline content was determined by DSC, the mica content by TGA, and the distribution and orientation of the mica platelets imposed by the injection molding process used by SEM and FESEM [3,4,17–20].

Aditionally, the preferential location of whatever interfacial agent in the inter-phase between the reinforcement (mica) and the matrix (iPP) was first proposed from DMA results [3,4,6,8,12,13,17–19] and further confirmed by SIRM (Synchrotron Infra-Red Microscopy) [20]. Preliminary studies allowed postulating the existence of chemical bonds between the mica and the interfacial agent [20]. This fact suggests that the iPP/aPP-pPBMA/mica system may be classified as a class II hybrid material (or at least as a mixture of I and II) wherein strong interactions occur [5].

Notwithstanding, the production of this kind of organic–inorganic hybrid materials implies a complex scenario depending on processing, composition, functionality, and emerging morphology. This permitted the researcher to obtain complex systems with consummate mastery at the different scales [3–5]. Therefore, a straightforward idea about what a complex system means consists of considering it as the one with many blocks able of exchanging stimuli between them and the surroundings depending on the contour conditions [3,4,21,22]. The former gives rise to behavior far from those expected from just the properties of the specific characteristics of the blocks (here iPP, mica, and aPP-*p*PBMA) that does not even give a glimpse of the behavior of the system itself [3–5,21,22]. Put differently, the complex systems require interaction effects between the blocks that overflow the expected by considering any additive effect of each of them [21,22]. In fact, these types of systems can be modeled by the so-called "agent-based models" [21,22]. Hence, the Box–Wilson surface response methodology for predicting the variation of the T_g with the combined effect of the mica and the interfacial agent content resembles these "agent-based models." Therefore, this permits the interpretation of the effect of both variables in the T_g variation with a physical sense [3,4,23,24]. In fact, the Box–Wilson response surface methodology can be considered as an "Agent-based Model" since it considers blocks (here named controlled factors), and a series of interaction terms helping to detect other effects of the materials rather than those of the controlled factors (mica and aPP-*p*PBMA). In our case, the external stimuli remained minimized since the processing method is the same for all the samples [3,4,23,24]. Moreover, the possible emergence of optimal coordinates in the experimental space scanned let discriminate the efficient and the non-efficient components Therefore, this article concerns the study and prediction of the glass transition temperature (T_g) looking to identify and interpret the combined and synergistic effect of the mineral reinforcement (mica), and the interfacial agent (aPP-*p*PBMA) on the variation and variability of the glass transition temperature of the polymer phase in the hybrid material as determined by Dynamic Mechanical Analysis [25,26].

2. Materials and Methods

2.1. Materials

The starting materials used were an isotactic polypropylene, iPP (ISPLEN 050) by Repsol (Madrid, Spain) with the following properties: $\rho = 0.90$ g/cm^3; $M_w = 334{,}400$; $M_n = 59{,}500$; $T_g = -13$ °C, and phlogopite mica platelets (KMg$_3$[Si$_3$AlO$_{10}$](OH)$_2$), Alsibronz® by BASF (Barcelona, Spain). The platy mica with density = 2.85 g/cm^3; specific surface BET = 1.5 m^2/g; average particle larger size = 79.8 µm was chosen by its demonstrated dimension stability, mean size and particle size distribution after and before the processing operations [3,4,14,17,18,20]. As interfacial agent, a grafted atactic polypropylene with 15% w/w (5×10^{-4} mol/g $_{polymer}$) *p*-phenylen-bis-maleamic acid attached groups (aPP-*p*PBMA) designed and obtained by the authors through a chemical modification process in the melt by using polymerization wastes as raw material. A fully detailed description of the process and the characterization procedures of the grafted polymer are fully described elsewhere [18]. Figure 1 shows a scheme of the chemical structure of the interfacial agent used.

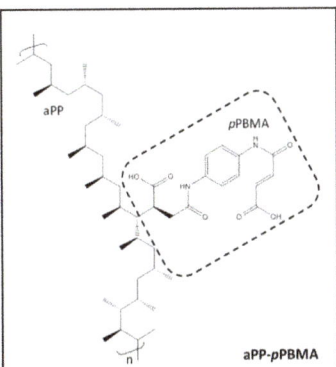

Figure 1. Chemical structure of the aPP-*p*PBMA interfacial agent used in this work.

2.2. Sample Preparation

The composites, according the doses conditions in Table 1, were compounded in a Rheomix 600 chamber connected to a Rheocord 90 (Haake, Barcelona, Spain) by the at a time addition of the platy mica and the interfacial agent (aPP-pPBMA) to the previously the molten iPP (190 °C). The interfacial agent (aPP-pPBMA) was incorporated to the iPP by just replacing the same amount of it in the compound. Therefore, once the torque was stabilized, after five minutes blending, the chamber was opened and the composites were cooled down into and ice bath.

Table 1. Experimental design and measured Glass Transition Temperature (T_g) according to the Box–Wilson experimental worksheet.

Exp	Controlled Factors *		Coded Factors		T_g (°C)	T_g (K)
	x_1 (%)	x_2 (%)	x_1	x_2		
E1	14.4	1.465	−1	−1	9.9	283.05
E2	35.6	1.465	1	−1	9.2	282.35
E3	14.4	8.535	−1	1	8.8	281.95
E4	35.6	8.535	1	1	6.7	279.85
E5	10.0	5.000	$-\sqrt{2}$	0	6.2	279.35
E6	40.0	5.000	$\sqrt{2}$	0	6.5	279.65
E7	25.0	0.001	0	$-\sqrt{2}$	10.9	284.05
E8	25.0	9.999	0	$\sqrt{2}$	8.4	281.55
E9	25.0	5.000	0	0	9.9	283.05
E10	25.0	5.000	0	0	10.0	283.15
E11	25.0	5.000	0	0	10.5	283.65
E12	25.0	5.000	0	0	9.9	283.05
E13	25.0	5.000	0	0	10.7	283.85

* x_1 = [Mica]; * x_2 = [aPP-pPBMA].

After that, once dried overnight at 25 °C, the hybrid material was milled to pellets and then injection molded at 200 °C in dog-bone type 1BA samples (ISO 527-2) by means of a Babyplast 6/6 micro-injection machine. From these, a series of prismatic samples (19.5 × 4 × 2 mm^3) shaped according the DMA test requirements, were obtained.

The real particle content in the composite was determined by thermo-gravimetric analysis (TGA) and the particle distribution by Field Emission Scanning Electron Microscopy (FESEM) and Environmental Scanning Electron Microscopy (ESEM) in previous articles by the authors [3,4,7,13,17].

2.3. Characterization

We used a dynamic mechanical analyzer, DMA, (METTLER DMA861, Madrid, Spain) under the tension mode to obtain the DMA spectra of all the compounds in Table 1. For such purpose, we followed the recommendations of ASTM D5026 standards. In this way, the dynamic mechanical parameters were measured within the range of linear viscous-elastic behavior of the material by considering 12N oscillating dynamic force applied at a fixed frequency (1 Hz) and 3 µm amplitude being the heating rate equal to 2 °C/min. The temperature was varied in the −40 to 60 °C interval. We use this frequency for better determining the interfacial effects. Moreover, both the rather low frequency and displacement applied to the samples in the DMA are due to avoid whatever nonlinear behaviour and any morphological changes provoked by eventual internal heat generation.

2.4. Mathematical Model

We employed Box–Wilson statistical experimental design (sDOE) to study and predict the glass transition variation of the iPP/aPP-pPBMA/mica organic–inorganic hybrid system. In essence. this methodology is in a central rotary composite design conssiting in (2^k + 2k + 1) experiments augmented with (2 + k) replicated runs in the central point coded as (0,0). Here, k is the number of the independent variables chosen (in our case mica and aPP-pPBMA) [23,24]. Therefore, and in order to obtain samples in the desired ranges, 0% up to 40% in mica content and 0% up to 10% for aPP-pPBMA in the organic–inorganic hybrids, an interval between 14.4% and 35.6%, and 1.465% and 8.535%, respectively must be considered, the latter is just the consequence of the factorial component coded as (−1, 1) in Box–Wilson methodology for the experimental space to be studied. Equally, the coded variable for the star points of the model is $\alpha = \sqrt{2}$ [23,24]. This coding, jointly with the uncoded (named as controlled) variables, have been included in Table 1. With this premise, the glass transition temperature measured for each experiment can be fitted, and thus a polynomial predicting (if adequate correlation is obtained) this property within the experimental range studied is obtained [23]. This information is included in Tables 2 and 3.

Table 2. Statistical Parameters and Coefficients of the Polynomials. (Polynomial Equation: $a_0 + a_1 \cdot x_1 + a_2 \cdot x_2 + a_3 \cdot x_1 \cdot x_2 + a_4 \cdot x_1^2 + a_5 \cdot x_2^2$) *.

	$<r^2>$ (%)	LF (%)	CF (%)	Linear Terms			Interaction Term	Quadratic Terms	
				a_0	a_1	a_2	a_3	a_4	a_5
T_g [K]	91.56	5.1	99.5	274.1	0.8024	0.07167	−0.009341	−0.011567	−0.009048

* x_1 = [Mica]; x_2 [aPP-pPBMA].

Table 3. Confidence coefficient (%) and t-values for the different terms of model obtained for T_g. (Polynomial Equation: $a_0 + a_1 \cdot x_1 + a_2 \cdot x_2 + a_3 \cdot x_1 \cdot x_2 + a_4 \cdot x_1^2 + a_5 \cdot x_2^2$) *.

	Linear Parameters		Interaction Parameter	Quadratic Parameters	
	x_1	x_2	$x_1 \cdot x_2$	x_1^2	x_2^2
T_g[K]	7.0 (99.9%)	0.25 (25.6%)	1.12 (68.9%)	7.5 (99.9%)	0.48 (36.7%)

* x_1 = [Mica]; x_2 [aPP-pPBMA].

3. Results and Discussion

Prior to discussing the DMA results, it is worth briefly and properly describing the scenario and possible interactions occurring when a polymer matrix (iPP in our case) hosts platelet-like reinforcement (here, mica), and an interfacial modifier is also present (aPP-pPBMA). This model was previously underlined by the authors elsewhere [3]. In fact, in the discussion on the glass transition results as DMA measured, we follow the theoretical approach that describes the scheme in Figure 2. This one primarily considers two diverse types of amorphous fraction in the iPP matrix [3]. One of them would be that allocated between the mica particle and the closest iPP crystalline phase surrounding it, while the other would be that interconnecting the crystals within the iPP crystalline macroaggregates. The latter ensures the matrix continuity as a whole and chiefly determines the iPP relaxation behavior well above the glass transition region. This is because it is primarily related to the processing and molding conditions. At the effects of the present discussion, it is the former that mainly influences the glass transition range. Regardless, it depends on the overall so-called "free amorphous" fraction within the net amorphous PP amount. This is mainly commanded (in these hybrid materials) by that fraction's highly constrained involvment in imbibing the inorganic particles.

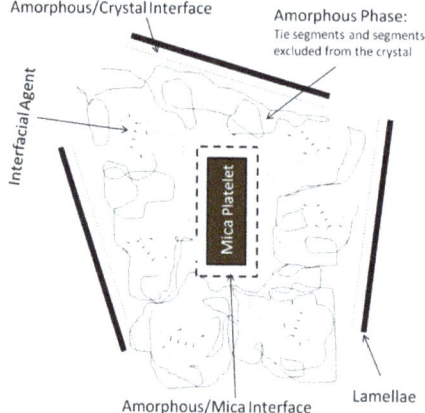

Figure 2. Scheme of the possible iPP/aPP-pPBMA/mica interactive scenario.

To begin, we can observe a simplification of the complex scenario related to the interactions occurring at the organic–inorganic interface. Initially, we appreciate that between the iPP crystalline phase (referred to as lamellae) and the inorganic phase, a series of zones can be defined. The first one between the crystal and the amorphous phase of the iPP matrix. Here, we can identify an amorphous/crystal interphase wherein the amorphous (not ordered) sequences of the isotactic polypropylene may be allocated jointly with some atactic sequences of iPP. Additionally, the inorganic phase (mica) is mandatorily embedded in the amorphous phase [3,4,6,7], and so, an amorphous/mica interface can be defined. Additionally, in between the iPP lamellae and the inorganic phase, the amorphous phase of the system, consisting of tie segments and other segments excluded from the crystal, is identified. Thus, and since the grafted groups are also excluded, from the crystalline domains, the interfacial agent, whatever its origin (isotactic or atactic) must be preferentially hosted in this area [3,4,7,8,11,12,17–20]. Even more, the interfacial agent used here (aPP-pPBMA), due to its amorphous origin, must be mandatorily allocated in this phase [3,4,7,8,12,13,17,18]. Hence, it results in clear that the presence of both the inorganic phase and the interfacial agent in the amorphous domains must disrupt this one. Consequently, it must influence the mobility of this amorphous phase responsible for the glass transition of the polymer. Thus, and depending on how crowded the zone is (by the presence of mica and aPP-pPBMA) and how intense the interactions are between them and the matrix, this transition must be necessarily affected [3,4,6,7].

In this sense, to replace a minor fraction of the polymer matrix with an interfacial agent greatly improves the interactions throughout the dynamic interface between the polymer and the reinforcement [3–5,8,12,13,17,18]. This implies the existence of a critical value depending on both the amount and the type of the interfacial agent used, and the processing history of the organic–inorganic hybrid material [7,17,18]. Notwithstanding, this is a core concept that is too often not considered in plenty of works in literature that do not pay attention to the processing and shaping operations. Therefore, these studies avoid considering that the properties of a polymer-based material strongly depend on how it has been conducted to the solid-state.

In a way, when studying the effect of the interfacial agent in an organic–inorganic hybrid composite consisting of changing the transport phenomena throughputs, we consider two possibilities. On one hand, the researcher can wield a constant amount of interface agent by varying the grafting level [11,12]. On the other, the interfacial agent may vary by keeping constant the graft percentage [3,4,7,17,18]. This last one represents the route we have adopted in the present study.

Additionally, the preferential orientation of the flow elements governing the preferential alignment of the platelets, their morphological variations, the particle size, and size distribution changes caused by the processing steps must not be assigned to modifications of the inter-phase (as many frequently occur in literature). In this sense, it is mandatory to indicate that the inorganic platelets used (mica) did not suffer significant changes in particle size and particle size distribution during processing [3,4,17–20]. This is the reason why a platy mica ($KMg_3[Si_3AlO_{10}](OH)_2$) providing a real reinforcement effect to the organic–inorganic material was chosen for this study [3,4,6,17–20]. Consequently, the real and precise mica content in the material must be ascertained. The latter in order to not falsely identify other interfacial effects than the interfacial agent and the mica caused by mere changes in the flow-dynamic of the system (and then hardly traceable). The latter was checked by the authors for the same experimental worksheet by means of TGA analysis [3,4,7,17,18].

3.1. Dynamic Mechanical Spectra: Determination of T_g

Here, we discuss the glass transition variations observed by the loss factor (tan δ) as determined by DMA for the iPP/Mica organic–inorganic hybrid material and the way this parameter is influenced by the combined effect of both the inorganic phase (mica) and the interfacial modifier used (aPP-pPBMA). Therefore, Figure 3 shows the evolution of the tan δ with temperature for all the samples of the Box–Wilson worksheet in Table 1. In this work, we have merely used the values for the glass transition temperature obtained from the tan δ plots in Figure 3. Typically, in the case of an iPP based composite this transition appears between −10 and 40 °C [3,4,7,11,12,17–19], being in our case between 6.2 °C (sample E5) up to 10.9 (sample E7), depending on the hybrid material formula (Table 1). It is significant to mention that this transition is related to the cooperative chain segments' motion on the "free" amorphous phase of the polymer wherein short-range diffusive chain motions takes place in spite of the low dissipation capability due to mere atomic vibration motions [3,4,7,11,12]. All the values of the glass transition temperature have been compiled in Table 1. Hence, as mentioned hitherto, a maximum difference of 4.6 °C between experiment E7 with 25% mica and a tiny amount of aPP-pPBMA ratio equal to 25/0; T_g = 10.9 °C and E5 with mica/aPP-pPBMA ratio equal to 10/5; T_g = 6.2 °C, suggesting that the combined effect of mica and aPP-pPBMA greatly affect the glass transition value. In the same sense, it is important to mention that in all the cases the T_g observed is higher to that of the neat iPP as determined by DMA under the same conditions (T_g = 4 °C) [2,3], and so we find a maximum difference between the neat iPP and the slightly modified compound (E7) of 6.9 °C, and a minimum of 2.2 °C if compared with the E5 sample, containing 10% of mica and 5% of aPP-pPBMA, suggesting that there is the combined effect of the reinforcement and the interfacial agent what affect the glass transition.

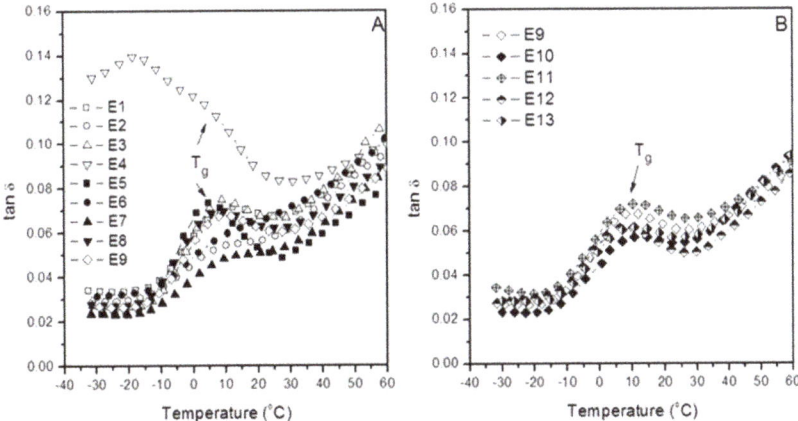

Figure 3. Evolution of the damp factor with temperature and glass transition for the indicated samples: (**A**) Central Rotary Composite Design Runs and (**B**) Central Point Replicated Runs.

Figure 3 shows the evolution of the loss factor with the temperature of each one of the compounds in Table 1. At a glance, we observe that the sample E4, with a 35.6/8.5 mica/aPP-pPBMA ratio, reveals a different pattern than the others. Thus, the latter evidence the abrupt increase in the dissipation capabilities of the iPP matrix in this organic–inorganic compound, similar to the reported for the 75/25 iPP/mica unmodified compound elsewhere [2,3,11,12]. The latter was explained on the basis of the well-known flatness of the mica particles rendering much amorphous iPP to imbibe them [3,4,12,13]. Therefore, in our case, the amount of mica is much higher (35.5%) but the amount of the amorphous character aPP-pPBMA is high enough as to provide sufficient extra amorphous phase aiding to embed the mica particles. Therefore, we found a similar dissipation mechanism in the case of a 25% unmodified mica compound [3,4,12,13] as in a 35.5/8.35 mica/aPP-pPBMA composite. On the contrary, just a tiny amount of aPP-pPBMA (sample E5) is enough to amend the interactions and so to alter the pattern regarding the unmodified iPP/Mica 75/25 unmodified composite [3,4,12,13]. The latter informs about the remarkably complex scenario that emerged from the possible interactions modeled in Figure 1. The mica particles act to disrupt the polymer bulk and consequently oblige a fraction of the polymer segments to be ordered [3,4,12,13]. Consequently, the amorphous region trapped at the iPP/Mica interface (Figure 1), which is coating the mica particles, is abruptly constrained. As follows, just a minor portion of the amorphous phase can become mobile [3,4,8,12,13]. Conversely, the presence of aPP-pPBMA in these regions of the organic–inorganic hybrid material may play a two-fold effect. On one hand, since its presence introduces supplementary amorphous material to the system, making more mobile one of the iPP phases, a decrease in the glass transition may be expected. On the other, the interactions between the pPMBA groups would be on the contrary sense by interacting with the mica domains. Thus, a complex scenario having influence in the ultimate value of the glass transition values looks to emerge. In fact, from the data and results in Table 1, it is impossible to discriminate at a glance about the effect of the components of the hybrid material on the values for the T_g. Under these auspices, the use of Box–Wilson methodology is revealed as a reliable tool to interpret the results.

3.2. Polynomial Fits and Analysis of Variance (ANOVA)

Table 1 compiled the values for T_g for the experimental design followed. The *w/w* amounts of mica and the interfacial agent (aPP-pPBMA), the controlled factors have been also listed. Thus, the T_g for each one the samples were fitted to a quadratic model by means of Box–Wilson surface response methodology [25] obtaining a polynomial describing the evolution of the glass transition temperature. Consequently, Table 2 compiles the terms of the polynomial obtained together to the lack of fit and the confidence factor coefficients for ANOVA (analysis of variance). Hence, we observe a value for equal to 91.56%, excellent for a quadratic model since values for this parameter higher than 75% are considered as good for these kinds of models [23,24]. Additionally, this Table 2 includes the "lack of fit" (LF), which is related to the percentage of pure error due to any factors overlooked by the model but significant enough in the final prediction of it. At this place, we obtain a value of LF equal to 5.1%, indicating that just this value may explain other factors ignored by the model. Likewise, the extraordinary value for the confidence factor (CF = 99.5%) indicates the accuracy and significance of the chosen independent variables chosen to model the T_g evolution of the iPP/aPP-pPBMA/Mica organic–inorganic hybrid material for the whole experimental space studied.

Moreover, the parameters included in Table 2 robustly confirm the possibility of studying this organic–inorganic hybrid system by means of the Box–Wilson predictions. In any case, it is important to check the limitations of the model. For such a purpose, the Figure 4 shows the scatter plots for the predicted versus the measured glass transition temperature. Thus, we can observe the excellent correlation between them.

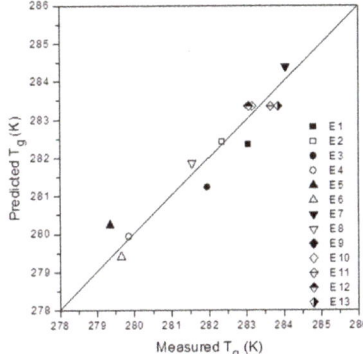

Figure 4. Measured versus predicted values for the glass transition temperature (T_g).

Hence, Table 3 compiles the confidence coefficient (%) and t-values for all the terms of the Box–Wilson polynomial relating the T_g with the composition of the hybrid material. At a glance, we notice the significance levels for each one of the parameters in the polynomial. Thus, by taking in mind that the more influence terms are those with t-values higher than two [23,24], we see that is the reinforcement that exercises more influence on the final T_g values; thus, the t-value = 7.0 for [Mica], and 7.2 for [Mica]^2 or confidence factors of 99.9% for both of them. The latter agrees with the above-mentioned point about the disruption capability of mica in order to immobilize the amorphous phase participating in the glass transition phenomena. However, these values for the interfacial agent in isolation are small (0.25 and 0.48 for the linear and the quadratic term, respectively), indicating that the interfacial agent in isolation has little influence in the final value of the T_g of the iPP matrix. However, the combined effect of mica and aPP-pPBMA appears to influence the final values of T_g since the t-value for the interaction term is 1.12, indicating that the confidence coefficient is close to 70%. The latter suggests that the effect of the interfacial agent is important enough to modulate the change in T_g mainly depending on the ability of mica to disrupt the amorphous phase of the iPP matrix [3–5,12,13].

3.3. Influence of the Composite Composition in the T_g as Determined by DMA Spectra

As mentioned in the previous sections, and prior to discussing the model predictions, we must make a series of remarks about the experimental data compiled in Table 1. It is not the case that the more interface agent and mica contents, the more change in property, which appears to indicate that the behavior of the iPP/Mica system is highly complex, and the effect of the presence of the interfacial agent is not evident. The latter implies this aspect may hardly be ascertained with just classical random experiments. Furthermore, the existence of critical values in the component concentration derived from the interactions between them rather than the effect of the components in isolation is what determines the overall behavior of the iPP glass transition evolution, as has been demonstrated for other mechanical properties [7,17–19] and also by DMA [3,4]. Notwithstanding, the evolution of the glass transition temperature of the iPP phase (in this organic–inorganic hybrid material) with the amount of filler and interfacial modifier can be studied and discussed on the basis of the Box–Wilson model predictions.

Thereinafter, Figure 5 shows the contour map of the glass transition of iPP as a function of the content of mica and interfacial agent. This map follows a rising ridge evolution, which means the existence of optimal coordinates in the experimental space scanned [23]. In our case, we observe quite close isolines for values below 15% and above 35% in mica content for whatever the amount of aPP-pPBMA, indicating there is a critical value for the mineral content, and the effect of increasing amounts of aPP-pPBMA is in the way to decrease the T_g. In fact, between 15% and 35% of mica, the separation of the isolines increases implying that for such reinforcement concentration the influence of the interfacial agent is less evident. Notwithstanding, the existence of a critical point is firmly established.

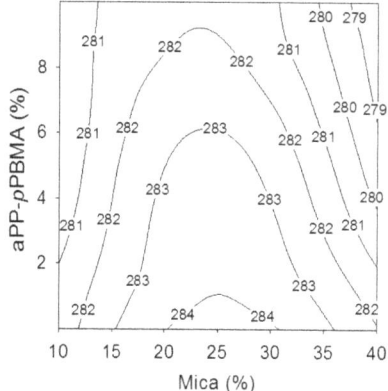

Figure 5. Evolution of T_g (K) with the Mica and aPP-pPBMA contents.

Additionally, and in order to facilitate a better comprehension of the Box–Wilson predictions, we have included two parametric plots wherein the existence of critical points is once more evidenced. Therefore, in Figure 6 we have represented the T_g versus the mica content in the hybrid material for the indicated amounts of aPP-pPBMA. Although all the curves follow a similar pattern with a maximum variation of T_g for the compound with 25% of mica independently to the aPP-pPBMA content, we observe that in all the cases the effect of aPP-pPBMA decreases the glass transition of the iPP phase, indicating the amorphous phase of the interfacial agent cooperates in hosting the mineral particles, and so a portion of the amorphous phase of the iPP matrix now becomes free as to participate in the local motions governing such transition, and consequently, the glass transition may occur at a lower temperature. What is clear is that the critical point in the system is the 25% of mica, for whatever the quantity of the interfacial agent used, meaning that at this point the reorganization capabilities of the crystal/amorphous balance of the iPP are optimized, and the role of aPP-pPBMA implies slight modulation of this effect.

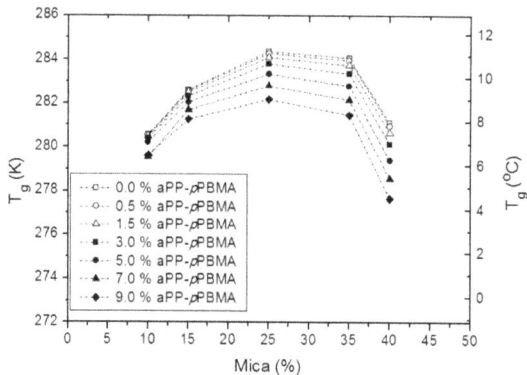

Figure 6. Evolution of T_g with the mica content at the indicated amount of interfacial agent.

In the same sense, Figure 7 plots the evolution of the glass transition of the iPP phase of the hybrid material with the aPP-pPBMA for different contents of mica. At a glance, we observe two families of curves with a similar slope to each other. One of the families, for values of mica below 25%, and other for values for mica 25% and higher. In this manner, the influence of the interfacial agent is in the way of decreasing the T_g of the iPP phase for whatever the amount of aPP-pPBMA used, but below 25% mica the decrease causes is low, due to the fact that the amount of amorphous phase of the iPP coating the mineral particles is not so constrained, and so the effect of the additional amorphous phase provided by the interfacial agent is not causing a considerable drop in the T_g values. However, values of mica equal to 25% and higher imply that this amorphous phase is therefore constrained that almost all of it is implied in covering the mineral and so, the difficulty of becoming mobile as to freely participate in the glass transition is high enough as to be sensitive to the incorporation of an additional amorphous phase coming from the interfacial agents aiding to coat the mineral flakes. At any rate, the more significant variations are found for the 25% mica compound, indicating that is the real critical point of the system, and so higher values than 25% imply lower T_g due to the different amorphous/crystal throughputs across the interphases.

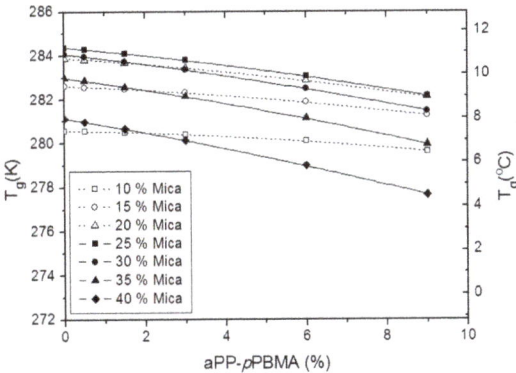

Figure 7. Evolution of T_g with the aPP-pPBMA content at the indicated amount of reinforcement.

To begin, the 25% mica coordinate is identified as the critical point of the system in the sense of increasing the glass transition value of the iPP phase. As follows, the interfacial agent plays a modulation effect on this transition. Nevertheless, the existence of this critical point maximizing this parameter is well evidenced. Furthermore, the fact that the latter is coincident to the determined for the ultimate properties (depending on the brittle to elastic capability) of the iPP/aPP-pPBMA/mica organic–inorganic hybrid material must be remarked [7,17–19].

4. Conclusions

The modulation role of aPP-pPBMA interfacial agent on the glass transition of the iPP phase of iPP/Mica composites has been determined by dynamic mechanical analysis (DMA). The use of the design of experiment methodologies (DOEs) has proved to be immensely useful in the study of systems wherein the interaction between the components determine the final property (the glass transition in this case). It is not the case that more of each component leads to a more outstanding level of interaction, which has been put into evidence once more. In fact, it is the existence of critical amounts of the components of the composite that governs the ultimate behavior of the system. In identifying the critical coordinates, the results obtained for the glass transition evolution agrees with those obtained for other ultimate properties. Therefore, this informs us of the linking of this threshold parameter (between the rigid-to-ductile behavior of the polymer matrix) with the ultimate behavior of the system. Thus, one emerging idea is that related to the effect of the reinforcement and the interfacial agent in the variation of the T_g of the polymers matrix. Here, we conclude that it depends on the interactions between the components in the sample jointly to the processing operations fingerprint. Consequently, those works in literature concluding that the effect on the T_g (increment or decrement of T_g) caused by any component of the hybrid material in isolation adquire non-sense. This applies consistently if a well-defined and controlled scenario (in terms of composition and processing history) is not considered.

Author Contributions: The two authors contributed equally to the study developed in the article. All authors have read and agreed to the published version of the manuscript.

Funding: The results discussed at present work were partially obtained under the auspices of the MAT 2000-1499 and MAT2013-47902-C2-1-R Research Projects.

Conflicts of Interest: The authors declare no conflict of interest.

References

1. Alemán, J.; Chadwick, A.V.; He, J.; Hess, M.; Horie, K.; Jones, R.G.; Kratochvíl, P.; Meisel, I.; Mita, I.; Moad, G.; et al. Definitions of terms relating to the structure and processing of sols, gels, networks, and inorganic–organic hybrid materials (IUPAC recommendations 2007). *Pure Appl. Chem.* **2007**, *79*, 1801–1829.
2. Pogrebnjak, A.D.; Beresnev, V.M. *Nanocoatings Nanosystems Nanotechnologies*; Bentham Books: Sharjah, UAE, 2012. [CrossRef]
3. García-Martínez, J.M.; Taranco, J.; Areso, S.; Collar, E.P. A DMA study of the interfacial changes on injection-molded iPP/mica composites modified by a *p*-phenylen-bis-maleamic acid grafted atactic polypropylene. *J. Appl. Polym. Sci.* **2017**, *134*, 45366. [CrossRef]
4. García-Martínez, J.M.; Areso, S.; Collar, E.P. The effect of a *p*-phenylen-bis-maleamic acid grafted atactic polypropylene interface agent on the dynamic-mechanical properties of iPP/mica composites measured at the polymer transitions. *Polym. Eng. Sci.* **2017**, *57*, 746–755.
5. Faustini, M.; Nicole, L.; Ruiz-Hitzky, E.; Sanchez, C. History of Organic-Inorganic Hybrid Materials: Prehistory, Art, Science, and Advanced Applications. *Adv. Funct. Mater.* **2018**, *28*, 1704158. [CrossRef]
6. Verbeeck, J.; Chistopher, M. Mica-Reinforced Polymer Composites. In *Polymer Composites. Vol I: Macro and Microcomposites*, 1st ed.; Thomas, S., Ed.; Wiley-VCH: Weinheim, Germany, 2012; pp. 673–714.
7. García-Martínez, J.M.; Areso, S.; Taranco, J.; Collar, E.P. The Role of a Novel *p*-phenylen-bis-maleamic acid grafted atactic polypropylene interfacial modifier in Polypropylene/Mica Composites as evidenced by Tensile Properties. *J. Appl. Polym. Sci.* **2009**, *113*, 3929–3943.
8. Collar, E.P.; Areso, S.; Taranco, J.; García-Martínez, J.M. Heterogeneous Materials based on Polypropylene. In *Polyolefin Blends*, 1st ed.; Nwabunma, D., Kyu, T., Eds.; Wiley-Interscience: Hoboken, NJ, USA, 2008; pp. 379–410.
9. Ciardelli, F.; Coiai, S.; Passaglia, E.; Pucci, A.; Ruggeri, G. Nanocomposites based on thermoplastic materials. *Polym. Int.* **2008**, *57*, 805–836.
10. Giannelis, E.P. Polymer layered silicate nanocomposites. *Adv. Mater.* **1998**, *8*, 29–35.
11. Karger-Kocsis, J. *Polypropylene, Structure, Blends and Composites*, 1st ed.; Chapman & Hall: London, UK, 1995.
12. García-Martínez, J.M.; Laguna, O.; Areso, S.; Collar, E.P. Polypropylene/Mica Composites Modified by Succinic Anhydride Grafted Atactic Polypropylene. A Thermal and Mechanical Study under Dynamical Conditions. *J. Appl. Polym. Sci.* **2001**, *81*, 625–636.
13. García-Martínez, J.M.; Laguna, O.; Areso, S.; Collar, E.P. A thermal and mechanical study under dynamical conditions of Polypropylene/Mica composites containing atactic polypropylene with succinil-fluoresceine grafted groups as interfacial modifier from the matrix side. *J. Polym. Sci. Polym. Phys.* **2000**, *38*, 1564–1574.
14. Karger-Kocsis, J.; Mahmood, H.; Pegoretti, A. Recent advances in fiber/matrix interphase engineering for polymer composited. *Prog. Polym. Sci.* **2015**, *73*, 1–43.
15. Pegoretti, A.; Karger-Kocsis, J. Interphase engineering in polymer composites: Challenging the devil . . . . *Express Polym. Lett.* **2015**, *9*, 838. [CrossRef]
16. Glotzer, S.C. Complex rules for soft systems. *Nat. Mater.* **2003**, *2*, 713–714. [PubMed]
17. García-Martínez, J.M.; Collar, E.P. Flexural behavior of PP/Mica composites interfacial modified by a *p*-phenylen-bis-maleamic acid grafted atactic polypropylene modifier obtained from industrial wastes. *J. Appl. Polym. Sci.* **2015**, *34*, 42437.
18. García-Martínez, J.M.; Cofrades, A.G.; Areso, S.; Collar, E.P. On the Chemical Modification Process of Atactic Polypropylene by *p*-Phenylen-bis-Maleamic Acid in the Melt. *J. Appl. Polym. Sci.* **2003**, *88*, 2202–2209.

19. García-Martínez, J.M.; Taranco, J.; Collar, E.P. Effect of a *p*-phenylen-bis-maleamic acid grafted atactic polypropylene interfacial agent on the impact properties of ipp/mica composites. *J. Appl. Polym. Sci.* **2017**, *134*, 44619.
20. Ellis, G.; Marco, C.; Gómez, M.A.; Collar, E.P.; García-Martínez, J.M. The study of heterogeneous polymer systems by synchrotron infrared microscopy. *J. Macromol. Sci. Phys.* **2004**, *B43*, 253–266.
21. Ottino, J.M. Complex Systems. *AIChe J.* **2003**, *49*, 292–299.
22. Ottino, J.M. Engineering Complex Systems. *Nature* **2004**, *427*, 399. [PubMed]
23. Box, G.E.P.; Hunter, W.G.; Hunter, J.S. Response surface methods. In *Statistics for Experimenters*, 1st ed.; Wiley & Sons: New York, NY, USA, 1978; pp. 510–539.
24. Fisher, R.A. *The Desing of Experiments*, 1st ed.; Hafner: New York, NY, USA, 1960.
25. McCrum, N.G.; Read, B.E.; Williams, G. *Anelastic and Dielectric Effects in Polymeric Solids*, 1st ed.; Wiley & Sons: London, UK, 1967.
26. McCrum, N.G.; Buckley, C.P.; Bucknall, C.B. *Principles of Polymer Engineering*, 1st ed.; Oxford University Press: New York, NY, USA, 1997.

Publisher's Note: MDPI stays neutral with regard to jurisdictional claims in published maps and institutional affiliations.

© 2020 by the authors. Licensee MDPI, Basel, Switzerland. This article is an open access article distributed under the terms and conditions of the Creative Commons Attribution (CC BY) license (http://creativecommons.org/licenses/by/4.0/).

Article

Regulation of Polyvinyl Alcohol/Sulfonated Nano-TiO$_2$ Hybrid Membranes Interface Promotes Diffusion Dialysis

Yuxia Liang [†], Xiaonan Huang [†], Lanzhong Yao, Ru Xia, Ming Cao, Qianqian Ge, Weibin Zhou, Jiasheng Qian, Jibin Miao * and Bin Wu *

Anhui Province Key Laboratory of Environment-Friendly Polymer Materials, School of Chemistry & Chemical Engineering, Anhui University, Hefei 230601, China; 13215692963@163.com (Y.L.); 19966505287@163.com (X.H.); ylz10001@163.com (L.Y.); xiarucn@sina.com (R.X.); cmcmycyc@163.com (M.C.); gqq@mail.ustc.edu.cn (Q.G.); zwb@ahu.edu.cn (W.Z.); qianjsh@ahu.edu.cn (J.Q.)
* Correspondence: lingxiaoyu1003@163.com (J.M.); lwbin@ahu.edu.cn (B.W.); Tel./Fax: +86-551-6386-1163 (J.M. & B.W.)
† These authors contributed equally to this work.

Abstract: It is important to emphasize that the adjustment of an organic–inorganic interfacial chemical environment plays an important role during the separation performance of composite materials. In this paper, a series of hybrid membranes were prepared by blending polyvinyl alcohol (PVA) solution and sulfonated nano-TiO$_2$ (SNT) suspension. The effects of different interfacial chemical surroundings on ions transfer were explored by regulating the dosage content of SNT. The as-prepared membranes exhibited high thermal and mechanical stability, with initial decomposition temperatures of 220–253 °C, tensile strengths of 31.5–53.4 MPa, and elongations at break of 74.5–146.0%. The membranes possessed moderate water uptake (WR) values of 90.9–101.7% and acceptable alkali resistances (swelling degrees were 187.2–206.5% and weight losses were 10.0–20.8%). The as-prepared membranes were used for the alkali recovery of a NaOH/Na$_2$WO$_4$ system via the diffusion dialysis process successfully. The results showed that the dialysis coefficients of OH$^-$ (U_{OH}) were in a range of 0.013–0.022 m/h, and separate factors (S) were in an acceptable range of 22–33. Sulfonic groups in the interfacial regions and –OH in the PVA main chains were both deemed to play corporate roles during the transport of Na$^+$ and OH$^-$.

Keywords: sulfonated; nano-TiO$_2$; diffusion dialysis; alkali recovery; assisted transport

Citation: Liang, Y.; Huang, X.; Yao, L.; Xia, R.; Cao, M.; Ge, Q.; Zhou, W.; Qian, J.; Miao, J.; Wu, B. Regulation of Polyvinyl Alcohol/Sulfonated Nano-TiO$_2$ Hybrid Membranes Interface Promotes Diffusion Dialysis. *Polymers* **2021**, *13*, 14. https://dx.doi.org/10.3390/polym13010014

Received: 4 September 2020
Accepted: 18 December 2020
Published: 23 December 2020

Publisher's Note: MDPI stays neutral with regard to jurisdictional claims in published maps and institutional affiliations.

Copyright: © 2020 by the authors. Licensee MDPI, Basel, Switzerland. This article is an open access article distributed under the terms and conditions of the Creative Commons Attribution (CC BY) license (https://creativecommons.org/licenses/by/4.0/).

1. Introduction

Diffusion dialysis (DD), which is driven by concentration gradient [1], is considered to be one of the most promising methods for alkaline waste water treatment as its spontaneous nature. In comparison with conventional separation processes, such as solvent extraction, precipitation, and distillation, the DD process exhibits significant superiority, including higher efficiency, low energy consumption, low installation and operating cost, stability and easiness for operation, and the environmentally friendly nature [2–4]. The core component of the DD process is the membrane. Therefore, membranes with excellent ion permeability and selectivity have been attracting increasing attention [5,6].

The alkaline DD process is not used as widely as it has been reported in acid recovery [7,8], which is due to the lower ion coefficients and selectivity. Recently, different efforts have been made to improve the separation performance of ion exchange membranes. These attempts could be broadly classified into three categories: new membrane-preparation methods from monomers that contain many ion exchange groups [5,9–11], composite membranes, and organic–inorganic hybrid membranes [12–15]. Among the rest, preparation of organic–inorganic hybrid membranes via blending or in situ methods draws the greatest attention because of their multiple functions—unique chemical reactivity, tunability of the organic polymer matrix, as well as the excellent mechanical and thermal stabilities of the inorganic backbone [3,16,17].

In recent years, organic–inorganic composites have been investigated for DD membranes with high performance [13,18]. Various nano-sized inorganic fillers such as silica [19], titania [20], zeolites [17], and montmorillonite [15,21] have already been used for the improvement of the performance of proton exchange membranes. The building of ion transfer channels between inorganic fillers and polymer matrices is the key to improve membrane separation performance [12]. However, it remains challenging to adjust the chemical environment of an interface in a composite membrane that can regulate the transportation efficiency of ions [13,14]. In our previous report [22–24], it was confirmed that functional groups in organic–inorganic interfacial regions could promote transport of ions. Therefore, sulfonated Nano-TiO$_2$ (SNT) and polyvinyl alcohol (PVA) were chosen as an inorganic filler and a polymer matrix to prepare hybrid membranes for alkaline DD in this research. The effects of –SO$_3^-$ from SNT and –OH from PVA on the performance of hybrid membranes were discussed preliminarily.

2. Materials and Methods

2.1. Materials

PVA was supplied by Sinopharm Chemical Regent Co., Ltd. (Shanghai, China). The average degree of polymerization was 1750 ± 50. Pre-weighed PVA was immersed in water and heated to around 100 °C and kept at 100 °C for 3 h. The homogeneous solution (5.0 wt %) was cooled to 60 °C before use.

Nano-TiO$_2$ powder was purchased from Nanosabz Co. Ltd. (Tehran, Iran), with average particle sizes of 30 nm, and was heated at 160 °C for 1 h before use. Sulfuric acid (H$_2$SO$_4$), glutaraldehyde (GA), acetone, and pure analytical toluene were purchased from Sinopharm Chemical Regent Co. Ltd. Pure analytical 1, 3-propanesultonewith was supplied by Shanghai Kang Ta chemical Co. Ltd. (Fengxian, Shanghai). Deionized water was used throughout.

2.2. Surface Modification of Nano-TiO$_2$

Sulfonation of nano-TiO$_2$ was illustrated in previous report [17,25,26]: 1 g of preheated nano-TiO$_2$ and 4.4 g 1, 3-propanesultone were added into 300 mL toluene under ultrasonic dispersion, and then, the mixture was stirred vigorously at 120 °C for 48 h. After that, the powders were soxhlet-extracted by acetone at 80 °C for 48 h. Finally, the samples were dried at 55 °C in a vacuum drying oven for 12 h to obtain SNT powders. The sulfonated process is presented in Scheme 1, and the sulfonation was confirmed by the FT-IR spectra, which was obtained by a NEXUS-870 (Thermo Fisher Scientific, Waltham, MA, USA) spectrometer.

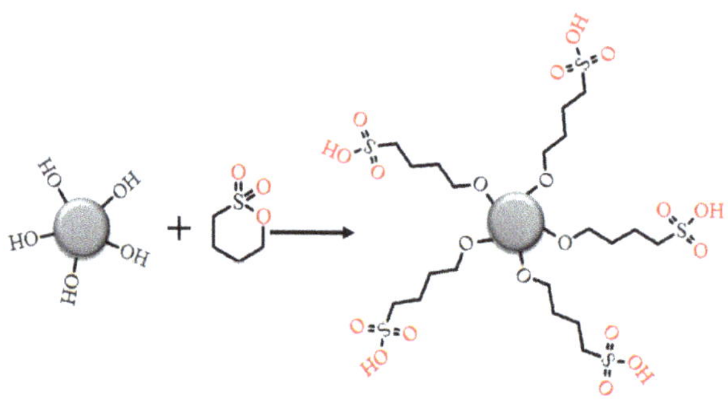

Scheme 1. Schematic diagram of sulfonation reaction of nano-TiO$_2$.

2.3. Preparation of the Hybrid Membranes

A preweighted GA aqueous solution (with a 0.5% mass ratio) was added to a 20 mL PVA aqueous solution under stirring, and the pH value was controlled at 5. After reaction for 35 min, SNT was added into a mixture of the prepared casting solutions and PVA with different mass ratios (0%, 1%, 3%, and 5%) under high-speed shear. Then, the solutions were casted onto clean glass plates and dried at room temperature for 48 h. The obtained membranes were dried from 60 °C to 130 °C at a rate of 10 °C/h and then kept at 130 °C for 5 h. The membranes were signed as 0%, 1%, 3%, and 5%, respectively, according to the dosage of SNT.

2.4. Characterization and Separation Performance of the As-Prepared Membranes

The microscopic structures and basic properties of the as-prepared membranes, including FT-IR (Thermo Scientific Nicolet iS10, Waltham, MA, USA), W_R, swelling degree, mass loss, mechanical and thermal properties (Instron 5967, Boston, MA, USA), and SEM and TEM images were measured, and the details could be seen in our previous reports [23,24]. The ion exchange capacities (IECs) of the as-prepared membranes were determined by the element analysis (Elementar Vario EL cube, Frankfurt, Germany) result of SNT. The DD test of the as-prepared membranes was detailed in our previous report [8,22], in which the effective area of membrane was 6 cm^2 and the temperature was kept at 25 °C. The solutions were stirred during the experiment, and the membrane samples were immersed into a feed solution for 2 h before testing.

3. Results

3.1. FTIR Spectra of Original Nano-TiO$_2$ and SNT and ATR-FTIR of the As-Prepared Membranes

The FTIR spectra of the original and sulfonated nano-TiO$_2$ are shown in Figure 1, in which the broad peak was observed in the region of 450–800 cm^{-1} corresponding to the Ti-O stretching vibration. The spectrum of the original nano-TiO$_2$ was characterized by the broad –OH stretching vibration and the bending vibration located at 3450 cm^{-1} and 1663 cm^{-1}, respectively. Compared with the original nano-TiO$_2$, the presentation of the S=O symmetric vibration peak at 1150–1210 cm^{-1} and the C–H adsorption peaks at 2920 cm^{-1} and 1380 cm^{-1} confirmed the achievement of SNT [27].

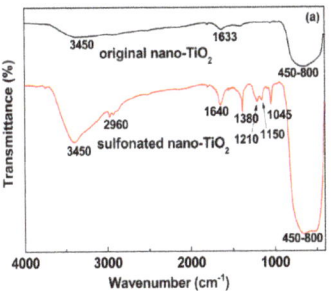
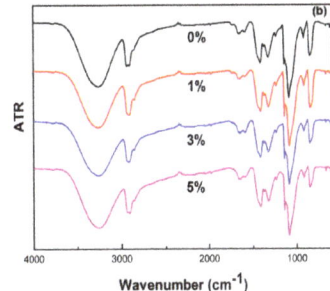

Figure 1. (a) FTIR spectra of the original nano-TiO$_2$ and the sulfonated nano-TiO$_2$. (b) ATR-FTIR spectra of the as-prepared membranes.

The ATR-FTIR spectra of the as-prepared membranes exhibited the similar trends, except for the peak between 1000 cm^{-1} and 1100 cm^{-1}, which could be assigned to C–O stretching and O–H bending vibrations coming from the PVA main chains. The detailed discussion is shown in Section 3.3.

3.2. IECs

The IECs of the as-prepared membranes were calculated from the results of a sulfur elemental analyzer (1.06%) and are shown in Table 1. The IEC values were in a range of

0–0.0157 mmol/g, which is much less than that of typical ion exchange membranes. This was mainly due to the lower sulfonated degree of SNT (about 1%). However, the introduction of sulfonic groups had much lower influence on the separation performance of membranes though the IECs of the as-prepared membranes than in previous reports [28,29].

Table 1. Theoretical ion exchange capacities (IECs) and thicknesses of the as-prepared membranes.

Membranes	0%	1%	3%	5%
IEC (mmol/g)	0	0.0033	0.0096	0.0157
Thickness (µm)	68	73	76	69

3.3. Water Uptake (W_R), Swelling Degree, and Mass Loss

The W_R results are shown in Figure 2, while the swelling degrees and the mass losses of as-prepared membranes are shown in Figure 3.

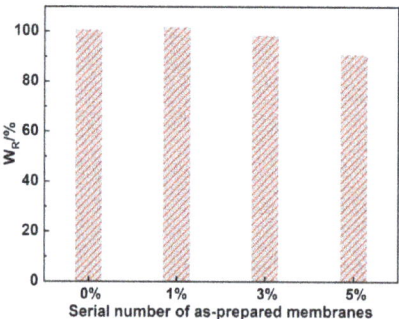

Figure 2. Water uptake (W_R) of the as-prepared membranes.

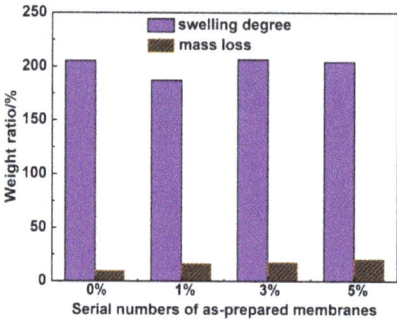

Figure 3. Swelling degrees and mass losses of the as-prepared membranes.

The W_R values of the membranes were in a range of 90.9–101.7%. The effect of the loading of SNT on the W_R values of membranes could be explained in two aspects: The hydrophilicity of hybrid membranes increased slightly, when the dosage of SNT was 1%, which was due to the incorporation of the hydrophilic –SO_3H and the well dispersion of the SNT particles; The values of W_R decreased obviously, while the dosage of SNT increased continuously. This could be attributed to the partial aggregation of the SNT, i.e., the SNT particles aggregated via the hydrogen bonds and this declined the number of dissociative –SO_3H. Meanwhile, considering the lower sulfonated degree of SNT, the increased relative mass ratio of SNT to PVA partially led to the decreasing W_R, when the dosage of SNT was more than 3%.

The swelling degrees of membranes in 2M NaOH at 65 °C were in a range of 187.2–206.5%, which was attributed to the degradation of crosslinked PVA chains. Membranes 1–5% exhibited slightly lower swelling degrees than that of membrane 0%, which indicated that the addition of SNT could restrict the movement of PVA chains. It could be observed obviously that the peak between 1000 cm^{-1} and 1100 cm^{-1} shifted to the higher wavenumber with the increasing dosage of SNT, which corresponded to those in a previous report [30]. The results indicated that incorporation of SNT was advantageous in the improvement of the swelling resistance of the hybrid membranes.

The mass loss of the membranes was mainly due to the dissociative PVA chains and resulted partly from the degradation of the crosslinking network under the attack of the hot alkali aqueous solution. All the membranes maintained integrity and original color after testing, while they turned brittle. This was due to the damage of the crosslinked structure caused by OH$^-$. The weight losses of the as-prepared membranes were in a range of 10.0–20.8%, compared with the finding in our previous report [23], and increased with the increasing dosage of SNT. This could be attributed to the introduction of sulfonic groups to the membrane matrix, that is, free H$^+$ from SNT facilitated the attack of OH$^-$ and the enhanced interfacial defect aggravated the erosion of the membranes. Therefore, the mass loss of the hybrid membranes increased with the increased SNT loading increment, and thus, an appropriate dosage of SNT was necessary.

3.4. Mechanical Properties

The tensile strength (TS) and elongation at break (E_b) of the as-prepared membranes are shown in Table 2. The TS values were in a range of 31.5–53.4 MPa, while the E_b values were in a range of 74.5–146.0%. The hybrid membranes possessed comparable mechanical strength and flexibility with those in our previous reports [22,23,31]. The strength and flexibility of the membranes declined, as the dosage content of SNT was enhanced, which indicated that the mechanical properties of the hybrid membranes were affected in an unconventional manner. Generally speaking, involvement of inorganic nanoparticles into a polymer matrix could improve mechanical properties of the composites due to the reduction of the free volume [23]. The unusual phenomena in this research could be explained as follows: –SO$_3$H group in the surface of SNT was easy to interact with the –OH in PVA main chains, which might lead to the rearrangement of the PVA molecular. Meanwhile, it could improve the chance to form caves between PVA main chain and SNT because of the similar hydrophily between –OH and –SO$_3$H [23]. Therefore, membranes 1%, 3%, and 5% showed declined mechanical properties compared with membrane 0%.

Table 2. Tensile strength (TS) and elongation at break (E_b) of membranes 0–5%.

Membranes	0%	1%	3%	5%
TS (MPa)	53.38	38.41	31.46	33.1
E_b (%)	146	108.28	90.02	74.48

3.5. Thermal Stabilities

TGA testing of the as-prepared membranes is shown in Figure 4. Since the membranes were heated at 130 °C, the weight loss before 130 °C could be neglected when determining the initial decomposition temperature (IDT).

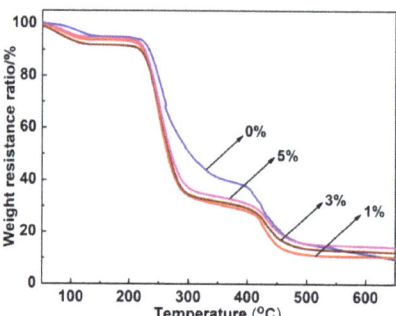

Figure 4. TGA results of the as-prepared membranes.

As can be see from Table 3, the IDT_1 values were the initial decomposition temperatures determined from thermograms, which were in a range of 220–253 °C, while the IDT_2 values were the second decomposition temperature of the other platform, which were in a range of 402–427 °C. With the increasing loading content of SNT, the IDT_1 and IDT_2 of hybrid membranes were higher than those of original one. The results demonstrated that introduction of SNT restricted PVA chains and enhanced the thermal stability of the hybrid membranes. The reasons were as follows: With the increasing loading content of SNT, the interaction between SNT and PVA molecules increased density of membranes and this was beneficial for the thermal stability of the membranes [27]. However, the excessive loading content of SNT could cause serious aggregation, which was disadvantage in the thermal stability of the as-prepared membranes. The crosslinking reaction was disadvantage in the dialysis of ions but advantage in the selectivity of the membranes. This would be further discussed in Section 3.7.

Table 3. Initial decomposition temperatures (IDTs) of the as-prepared membranes.

Membrane	0%	1%	3%	5%
IDT_1 (°C)	220	247	253	250
IDT_2 (°C)	402	425	431	427

IDT_1 was the initial decomposition temperature determined from thermograms. IDT_2 was the second decomposition temperature of the second platform.

3.6. Microscopic Morphologies

The cross-sectional SEM pictures of the as-prepared membranes are shown in Figure 5. The hybrid membranes exhibited obvious phase interface, while the original one showed a smooth broken surface. SNT dispersed uniformly in hybrid membranes, and there were not obvious caves or structure defects in the low-magnification SEM images, even when the loading content of SNT was 5%, which indicated well compatibility between the two phases. The SNT showed some slight aggregation when the dosage content was 5%, which could be attributed to the formation of H-bonding by $-SO_3H$ in the surface of SNT [23]. Nonetheless, there were obvious little cracks in the high-magnification images, especially in images 3%-1 and 5%-1. This indicated that the incorporation of SNT affected the arrangement of the PVA chains and the rearrangement of the PVA chains enhanced the chance of formation of structural caves. The SEM images agreed with the analytical results of the thermal and mechanical results as well.

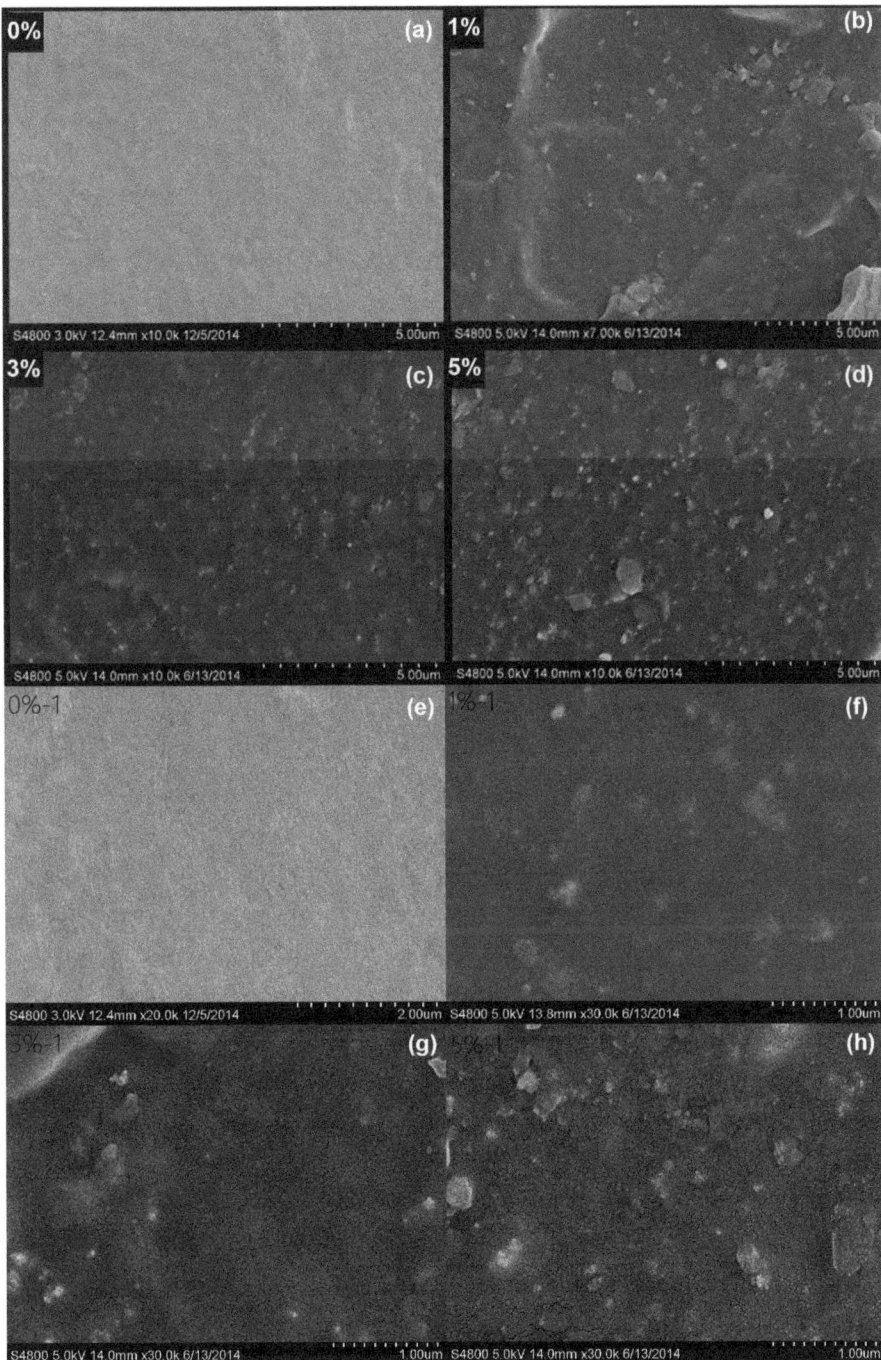

Figure 5. Cross-sectional SEM images of the as-prepared membranes: (**a–d**) 0–5% with low magnification; (**e–h**) 0%-1–5%-1 with high magnification.

To further detect the dispersion of SNT in the PVA matrix, the TEM images of samples 3% and 5% were taken, and the results are shown in Figure 6, from which we observed that SNT dispersed uniformly in membrane 3% while aggregated in membrane 5%. This agreed with the observations in the SEM images.

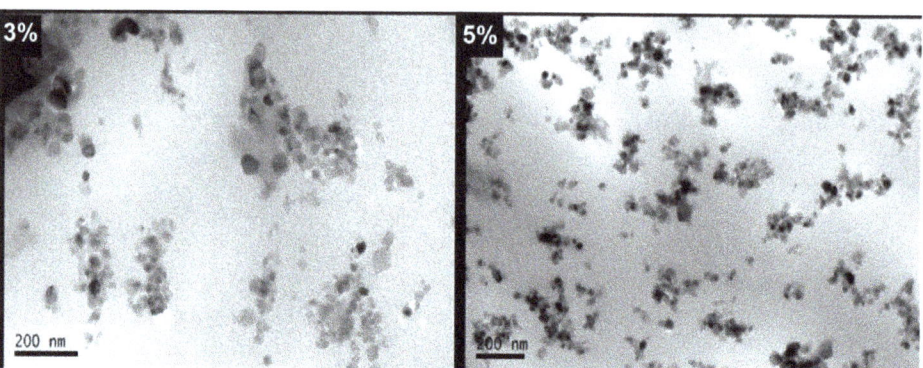

Figure 6. TEM photographs of membranes 3% and 5%.

3.7. Separation Performance

3.7.1. Dialysis Coefficients (U_{OH})

The dialysis coefficients of OH$^-$ (U_{OH}) are shown in Figure 7. The U_{OH} values were in a range of 0.013–0.022 m/h, which was comparable with previous reports [12,14]. As shown in Figure 7, U_{OH} values increased significantly and then decreased slightly with the increasing loading content of SNT. The results indicated that the incorporation of sulfonic groups was beneficial to the transport of OH$^-$. It is known that the IEC plays an important role during ion transport because cationic ions could easily traverse through the membrane via electrostatic attraction [1]. Nonetheless, compared with typical cationic ion exchange membranes (CIEMs) [5,6], the hybrid membranes possessed two characters: IECs were far lower than those of typical CIEMs and the functional groups were in the organic–inorganic interfacial regions. Therefore, it was concluded that ion exchange groups in the organic–inorganic interfacial regions could facilitate the transport of ions. This could be explained in two aspects: On the one hand, the larger interfacial space was easier for transport of Na$^+$ via electrostatic attraction; On the other hand, hydroxyl in the PVA main chains could promote transport of OH$^-$ through the hydrogen bond [22]. The two factors played synergistic roles on the ion transport during the DD process and enhanced the U_{OH} to 0.022 m/h, which was nearly twice as much as the U_{OH} of a pure PVA membrane. The transport schematic diagram is shown in Figure 8. However, aggregation appeared, and membranes separation performance declined at the highest loading of 5%, which was in step with the results of SEM and TEM results. The reasons for this were as follows: –SO$_3$H in the surface of SNT formed the transported channel for Na$^+$ in the larger interfacial space via electrostatic attraction and this decided the separation performance of the hybrid membranes. The SNT itself aggregated via the H-bonding between –SO$_3$H, and this decreased the dissociative number of –SO$_3$H. Thus, the transport of Na$^+$ was delayed, and the separation of membrane declined.

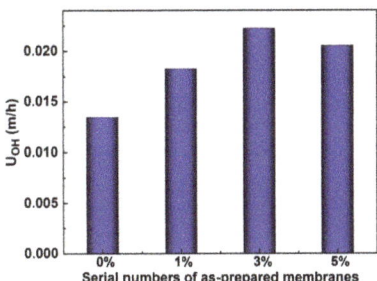

Figure 7. Dialysis coefficients of OH^- (U_{OH}).

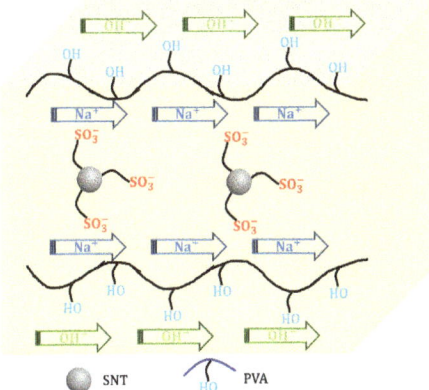

Figure 8. Transport schematic diagram of Na^+ and OH^-.

3.7.2. Separation Factors (S)

The S values of the as-prepared membranes are shown in Figure 9, from which higher S values of the hybrid membranes than that of the original one were observed. The results indicated that the incorporation of SNT was beneficial to the selectivity of membranes. The S values were in an acceptable range of 22–33, which was lower than that of SPPO-based hybrid cation exchange membranes [6,9].

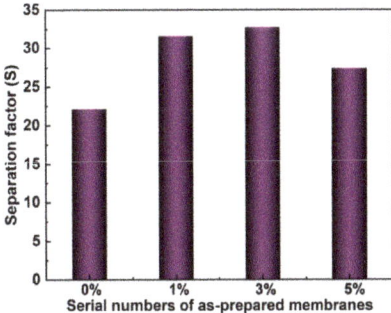

Figure 9. Separation factors (S) of the as-prepared membranes.

The S values of the hybrid membranes increased as the loading content of SNT increased and decreased slightly while the loading content of SNT reached 5%. This could be explained as follows: All the membranes possessed high density after thermal treatment, and this would make it difficult for the transport of WO_4^{2-} because of its bigger volume and higher valence state. On the contrary, transport of OH^- was less affected because of its smaller volume and lower valence state than WO_4^{2-}. More important, hydroxyl in the PVA main chains provided the assisted transport of OH^- via hydrogen bonding (seen in Figure 8). Meanwhile, WO_4^{2-} suffered larger electrostatic repulsion while transport through the membrane, which was due to its multivalent property and larger volume.

Therefore, the incorporation of SNT to PVA matrix could enhance U_{OH} and S simultaneously, and this might be one of the candidates to break down the "tradeoff" effect between ion flux and selectivity. Compared with the findings in our previous report [8,22–24], SNT exhibited higher stability in an alkaline system than in silica, and the hybrid membranes showed excellent properties under lower dosage (less than 5%). Meanwhile, the sulfonation of nano-TiO_2 was easy to carry out, and the blending method was easy to commercialize. However, aggregation appeared with excessive dosage of SNT, which declined the performance of membranes as discussed in Section 3.7.1. Therefore, it was important to enhance the dispersion abilities of SNT in the polymer matrix under some additional technology (such as ultrasonic dispersion, high-speed shear and in situ preparation). Thus, suitable loading content and multiple-function surface modification of SNT could help obtain membranes with the best performance. In this system, the hybrid membranes exhibited optimal performance, when the loading content of SNT was 3%, with U_{OH} and S were 0.022 m/h and 33, respectively.

4. Conclusions

PVA/nano-TiO_2 hybrid membranes have been prepared by bending a precrosslinked PVA solution and an SNT suspension. The SEM and TEM images confirmed the good compatibility between these two phases. The as-prepared membranes were of good hydrophilicity and moderate alkali resistance, with W_R values of 90.9–101.7% and weight losses of 10.0–20.8%. The results of TGA and mechanical tests indicated that the as-prepared membranes were of thermal and mechanical stability with initial decomposition temperatures of higher than 220 °C, tensile strengths (TS) of 31.5–53.4 MPa, and elongation at break (E_b) of 74.5–146.0%. The as-prepared membranes were applied to recover alkali from the $NaOH/Na_2WO_4$ system via the DD process successfully, and dialysis coefficients of OH^- (U_{OH}) and separation factor (S) values were in a range of 0.013–0.022 m/h and 22–33, respectively. The sulfonic groups in the organic–inorganic interface of the hybrid membranes and –OH from PVA main chains were deemed to play important roles during the DD process: Na^+ transported through the main channels made up of $-SO_3^-$, while OH^- transported through the assisted channels made up of –OH from PVA. Thus, ion flux and selectivity could enhance simultaneously by the incorporation of SNT. The membrane exhibited an optimal performance when the loading content of SNT was 3%, with U_{OH} of 0.022 m/h and S of 33.

Author Contributions: Funding acquisition, J.Q.; investigation, R.X., J.Q. and B.W.; methodology, Q.G. and W.Z.; resources, L.Y.; supervision, J.M.; writing of an original draft, Y.L. and X.H.; writing of review and editing, M.C. All authors have read and agreed to the published version of the manuscript.

Funding: This work was supported by the National Natural Science Foundation of China (Nos. 51973002, 21606001, 21606215, and 21808001), the Anhui Provincial Natural Science Foundation (No. 1708085QE117), and the Doctoral Scientific Research Startup Foundation of Anhui University (No. J01003213). Particularly, the financial supports from the University Synergy Innovation Program of Anhui Province (GXXT-2019-001 and GXXT-2019-030), Collaborative Innovation Center for Petrochemical New Material and Institute of High-Performance Rubber Materials & Products of Anhui Province were appreciated.

Conflicts of Interest: The authors declare no conflict of interest.

References

1. Luo, J.; Wu, C.; Xu, T.; Wu, Y. Diffusion dialysis-concept, principle and applications. *J. Membr. Sci.* **2011**, *366*, 1–16. [CrossRef]
2. Naik, N.S.; Padaki, M.; Deon, S.; Murthy, D.H.K. Novel poly (ionic liquid)-based anion exchange membranes for efficient and rapid acid recovery from industrial waste. *Chem. Eng. J.* **2020**, *401*. [CrossRef]
3. Hong, J.G.; Chen, Y. Nanocomposite reverse electrodialysis (RED) ion-exchange membranes for salinity gradient power generation. *J. Membr. Sci.* **2014**, *460*, 139–147. [CrossRef]
4. Yadav, V.; Raj, S.K.; Rathod, N.H.; Kulshrestha, V. Polysulfone/graphene quantum dots composite anion exchange membrane for acid recovery by diffusion dialysis. *J. Membr. Sci.* **2020**, *611*. [CrossRef]
5. Xiao, X.; Wu, C.; Cui, P.; Luo, J.; Wu, Y.; Xu, T. Cation exchange hybrid membranes from SPPO and multi-alkoxy silicon copolymer: Preparation, properties and diffusion dialysis performances for sodium hydroxide recovery. *J. Membr. Sci.* **2011**, *379*, 112–120. [CrossRef]
6. Liu, R.; Wu, L.; Pan, J.; Jiang, C.; Xu, T. Diffusion dialysis membranes with semi-interpenetrating network for alkali recovery. *J. Membr. Sci.* **2014**, *451*, 18–23. [CrossRef]
7. Yadav, V.; Rajput, A.; Kulshrestha, V. Sulfonated Poly(ether sulfone) based sulfonated molybdenum sulfide composite membranes and their applications in salt removal and alkali recovery. *J. Membr. Sci.* **2020**, *603*. [CrossRef]
8. Wang, C.W.; Liang, Y.X.; Miao, J.B.; Wu, B.; Hossain, M.M.; Cao, M.; Ge, Q.Q.; Su, L.F.; Zheng, Z.Z.; Yang, B.; et al. Preparation and properties of polyvinyl alcohol (PVA)/mesoporous silica supported phosphotungstic acid (MS-HPW) hybrid membranes for alkali recovery. *J. Membr. Sci.* **2019**, *592*. [CrossRef]
9. Wang, C.; Wu, C.; Wu, Y.; Gu, J.; Xu, T. Polyelectrolyte complex/PVA membranes for diffusion dialysis. *J. Hazard. Mater.* **2013**, *261*, 114–122. [CrossRef]
10. Wu, Y.; Luo, J.; Zhao, L.; Zhang, G.; Wu, C.; Xu, T. QPPO/PVA anion exchange hybrid membranes from double crosslinking agents for acid recovery. *J. Membr. Sci.* **2013**, *428*, 95–103. [CrossRef]
11. Xiao, X.L.; Shehzad, M.A.; Yasmin, A.; Ge, Z.J.; Liang, X.; Sheng, F.M.; Ji, W.G.; Ge, X.L.; Wu, L.; Xu, T.W. Anion permselective membranes with chemically-bound carboxylic polymer layer for fast anion separation. *J. Membr. Sci.* **2020**, *614*. [CrossRef]
12. Wu, Y.; Lin, H.; Zhang, G.; Xu, T.; Wu, C. Non-charged PVA/SiO$_2$ hybrid membranes for potential application in diffusion dialysis. *Sep. Purif. Technol.* **2013**, *118*, 359–368. [CrossRef]
13. Wu, Y.; Hao, J.; Wu, C.; Mao, F.; Xu, T. Cation exchange PVA/SPPO/SiO$_2$ membranes with double organic phases for alkali recovery. *J. Membr. Sci.* **2012**, *424*, 383–391. [CrossRef]
14. Hao, J.; Gong, M.; Wu, Y.; Wu, C.; Luo, J.; Xu, T. Alkali recovery using PVA/SiO$_2$ cation exchange membranes with different -COOH contents. *J. Hazard. Mater.* **2013**, *244*, 348–356. [CrossRef]
15. Lin, Y.F.; Yen, C.Y.; Hung, C.H.; Hsiao, Y.H.; Ma, C.C.M. A novel composite membranes based on sulfonated montmorillonite modified Nafion® for DMFCs. *J. Power Sources* **2007**, *168*, 162–166. [CrossRef]
16. Kim, Y.; Choi, Y.; Kim, H.K.; Lee, J.S. New sulfonic acid moiety grafted on montmorillonite as filler of organic-inorganic composite membrane for non-humidified proton-exchange membrane fuel cells. *J. Power Sources* **2010**, *195*, 4653–4659. [CrossRef]
17. Yu, D.M.; Yoon, Y.J.; Kim, T.H.; Lee, J.Y.; Hong, Y.T. Sulfonated poly(arylene ether sulfone)/sulfonated zeolite composite membrane for high temperature proton exchange membrane fuel cells. *Solid State Ionics* **2013**, *233*, 55–61. [CrossRef]
18. Luo, J.; Wu, C.; Wu, Y.; Xu, T. Diffusion dialysis of hydrochloride acid at different temperatures using PPO/SiO$_2$ hybrid anion exchange membranes. *J. Membr. Sci.* **2010**, *347*, 240–249. [CrossRef]
19. Choi, Y.; Kim, Y.; Kim, H.K.; Lee, J.S. Direct synthesis of sulfonated mesoporous silica as inorganic fillers of proton-conducting organic-inorganic composite membranes. *J. Membr. Sci.* **2010**, *357*, 199–205. [CrossRef]
20. Yang, C.C. Synthesis and characterization of the cross-linked PVA/TiO$_2$ composite polymer membrane for alkaline DMFC. *J. Membr. Sci.* **2007**, *288*, 51–60. [CrossRef]
21. Kim, Y.; Lee, J.S.; Rhee, C.H.; Kim, H.K.; Chang, H. Montmorillonite functionalized with perfluorinated sulfonic acid for proton-conducting organic-inorganic composite membranes. *J. Power Sources* **2006**, *162*, 180–185. [CrossRef]
22. Miao, J.; Yao, L.; Yang, Z.; Pan, J.; Qian, J.; Xu, T. Sulfonated poly(2,6-dimethyl-1,4-phenyleneoxide)/nano silica hybrid membranes for alkali recovery via diffusion dialysis. *Sep. Purif. Technol.* **2015**, *141*, 307–313. [CrossRef]
23. Miao, J.; Li, X.; Yang, X.Z.; Jiang, C.; Qian, J.; Xu, T. Hybrid membranes from sulphonated poly (2, 6-dimethyl-1, 4-phenylene oxide) and sulphonated nano silica for alkali recovery. *J. Membr. Sci.* **2016**, *498*, 201–207. [CrossRef]
24. Li, X.; Miao, J.; Xia, R.; Yang, B.; Chen, P.; Cao, M.; Qian, J. Preparation and properties of sulfonated poly (2, 6-dimethyl-1, 4-phenyleneoxide)/mesoporous silica hybrid membranes for alkali recovery. *Micropor. Mesopor. Mater.* **2016**, *236*, 48–53. [CrossRef]
25. Fatyeyeva, K.; Bigarré, J.; Blondel, B.; Galiano, H.; Gaud, D.; Lecardeur, M.; Poncin-Epaillard, F. Grafting of p-styrene sulfonate and 1,3-propane sultone onto laponite for proton exchange membrane fuel cell application. *J. Membr. Sci.* **2011**, *366*, 33–42. [CrossRef]
26. Munakata, H.; Chiba, H.; Kanamura, K. Enhancement on proton conductivity of inorganic-organic composite electrolyte membrane by addition of sulfonic acid group. *Solid State Ionics* **2005**, *176*, 2445–2450. [CrossRef]
27. Kabiri, K.; Zohuriaan-Mehr, M.J.; Mirzadeh, H.; Kheirabadi, M. Solvent-, ion- and pH-specific swelling of poly(2-acrylamido-2-methylpropane sulfonic acid) superabsorbing gels. *J. Polym. Res.* **2010**, *17*, 203–212. [CrossRef]

28. Hao, J.; Wu, Y.; Ran, J.; Wu, B.; Xu, T. A simple and green preparation of PVA-based cation exchange hybrid membranes for alkali recovery. *J. Membr. Sci.* **2013**, *433*, 10–16. [CrossRef]
29. Wu, Y.; Wu, C.; Li, Y.; Xu, T.; Fu, Y. PVA-silica anion-exchange hybrid membranes prepared through a copolymer crosslinking agent. *J. Membr. Sci.* **2010**, *350*, 322–332. [CrossRef]
30. Vinod, V.T.P.; Nguyen, N.H.A.; Sevcu, A.; Cernik, M. Fabrication, Characterization, and Antibacterial Properties of Electrospun Membrane Composed of Gum Karaya, Polyvinyl Alcohol, and Silver Nanoparticles. *J. Nanomater.* **2015**. [CrossRef]
31. Chong, F.; Wang, C.; Miao, J.; Xia, R.; Cao, M.; Chen, P.; Yang, B.; Zhou, W.; Qian, J. Preparation and properties of cation-exchange membranes based on commercial chlorosulfonated polyethylene (CSM) for diffusion dialysis. *J. Taiwan Inst. Chem. Eng.* **2017**, *78*, 561–565. [CrossRef]

MDPI
St. Alban-Anlage 66
4052 Basel
Switzerland
Tel. +41 61 683 77 34
Fax +41 61 302 89 18
www.mdpi.com

Polymers Editorial Office
E-mail: polymers@mdpi.com
www.mdpi.com/journal/polymers

www.ingramcontent.com/pod-product-compliance
Lightning Source LLC
LaVergne TN
LVHW070704100526
838202LV00013B/1031